Lecture Notes in Mathematics

Edited by A. Dold and B. Eckmann

1148

Mark A. Kon

Probability Distributions in Quantum Statistical Mechanics

Springer-Verlag
Berlin Heidelberg New York Tokyo

Author

Mark A. Kon
Department of Mathematics, Boston University
111 Cummington Street
Boston, MA 02215, USA

Mathematics Subject Classification (1980): 82A15, 60E07, 60G60, 46L60

ISBN 3-540-15690-9 Springer-Verlag Berlin Heidelberg New York Tokyo
ISBN 0-387-15690-9 Springer-Verlag New York Heidelberg Berlin Tokyo

Printing and binding: Beltz Offsetdruck, Hemsbach/Bergstr.
2146/3140-543210

PREFACE

The purpose of this work is twofold: to provide a rigorous mathematical foundation for study of the probability distributions of observables in quantum statistical mechanics, and to apply the theory to examples of physical interest. Although the work will primarily interest mathematicians and mathematical physicists, I believe that results of purely physical interest (and at least one rather surprising result) are here as well. Indeed, some (§9.5) have been applied (see [JKS]) to study a model of the effect of angular momentum on the frequency distribution of the cosmic background radiation. It is somewhat incongruous that in the half century since the development of quantum statistics, the questions of probability distributions in so probabilistic a theory have been addressed so seldom. Credit is due to the Soviet mathematician Y.A. Khinchin, whose *Mathematical Foundations of Quantum Statistics* was the first comprehensive work (to my knowledge) to address the subject.

Chapters 7 and 8 are a digression into probability theory whose physical applications appear in Chapter 9. These chapters may be read independently for their probabilistic content. I have tried wherever possible to make the functional analytic and operator theoretic content independent of the probabilistic content, to make it accessible to a larger group of mathematicians (and hopefully physicists).

My thanks go to I.E. Segal, whose ideas initiated this work and whose work has provided many of the results needed to draw up the framework developed here. My thanks go also to Thomas Orowan, who saw the input and revision of this manuscript, using TEX, from beginning to end; his work was invariably fast and reliable. Finally I would like to express my appreciation to the Laboratory for Computer Science at M.I.T., on whose DEC 10 computer this manuscript was compiled, revised, and edited.

CONTENTS

CHAPTER 1

INTRODUCTION

§1.1 Purposes and Background

The most general mathematical description of an equilibrium quantum system is in its density operator, which contains all information relevant to the probability distributions of associated observables. Let S be such a system, with Hamiltonian H calculated in a reference frame with respect to which S has zero mean linear and angular momentum. Let $\beta = \frac{1}{kT}$ be the inverse temperature of S (T is temperature and k is Boltzmann's constant). If the operator $e^{-\beta H}$ is trace class the appropriate density operator for S is

$$\rho = \frac{e^{-\beta H}}{\text{tr } e^{-\beta H}}.$$

Our general purpose is to obtain probability distributions of observables in S from spectral properties of H. If H has very dense point spectrum, distributions depend on it only through an appropriate spectral measure; if the spectrum has a continuous component then $e^{-\beta H}$ has infinite trace, so that the associated density operator and probability distributions must again be defined via spectral measures. We will study how distributions are determined by spectral measures μ and apply the resulting theory to the (continuous spectrum) invariant relativistic Hamiltonian in Minkowski space and to the (discrete spectrum) invariant Hamiltonian in a spherical geometry (Einstein space), to derive probability distributions of certain important observables.

In some systems with continuous spectrum, a natural μ obtains through a net $\{H_\epsilon\}_{\epsilon > 0}$ of operators with pure point spectrum which in an appropriate sense approximates H. In order that subsequent conclusions be well-founded, it must be required that μ be independent of the choice of $\{H_\epsilon\}_{\epsilon > 0}$, within some class of physically appropriate or "natural" nets. For example, if A is an elliptic operator on a non-compact Riemannian manifold M, a natural class arises in approximating M by large compact manifolds. The well-developed theory of spectral asymptotics of pseudodifferential operators is useful here (see, e.g., [H], [See]). The procedure of infinite volume limits has also been studied and developed in Schrödinger theory (see [Si, Section C6]). In physical applications such as to the Planck law for photons, this approximation procedure is appropriate since distributions must be localized spatially, as well as with respect to wave propagation vector.

In the systems we consider, many interesting observables are sums of independent ones indexed by the spectrum of a (maximal) commuting set of observables. Thus in cases of continuous and asymptotically continuous spectrum, the notion of sum of independent random variables is very naturally replaced by that of an integral (over the spectrum of observables), defined in a way completely analogous to the Riemann integral. The theory of such integrals and associated central limit theorems will be developed (Chapters 3-5), and then applied to particular random variables of interest.

A more comprehensive and abstract theory will be studied in Chapters 7-9. The integration procedure will be part of a more complete Lebesgue integration theory for random variable-valued functions. Connections will be made with the theory of random distributions and purely random fields [V,R1].

The related work in this area has been done primarily on R^n, with Lebesgue measure. General information on random distributions is contained in [GV]. Multi-dimensional white noise was introduced in [Che], motivated by study of so-called Lévy Brownian motion in [Lé]; the topic was further developed in [Mc]. A comprehensive theory of generalized random fields was introduced by Malchan [M], using the notion of biorthogonality of random distributions on R^n.

The non-linearity of the Lebesgue integral in Chapter 7 is not essential, since it is equivalent to an ordinary (linear) Lebesgue integral of a distribution valued random variable field, and equivalently an integral over a space of logarithms of characteristic functions. Thus the integral is a simple and fundamental object. The probabilistic content is itself novel and (hopefully) interesting, and these chapters have been written largely as a semi-autonomous part of the monograph. Probability and statistical mechanics are re-joined in Chapter 9.

There are two physical implications of this work which warrant attention. The first is that non-normally distributed observables (such as photon number) can arise in physically attainable situations, namely those in which spectral density is non-vanishing near zero energy; the latter occurs in systems which are approximately one-dimensional, for example, in optic fibers or wave guides. The second is that within the class of models in which the density operator depends only on the Hamiltonian, the "blackbody spectrum" in a large-scale equilibrium system of photons admits an energy density spectrum which follows a classical Planck law, whose specific form depends only on basic geometry. This work provides a rigorous basis for the study of blackbody radiation in general spatial geometries, specifically the spherical geometry of Einstein space. This is a model of the physical universe (see [Se2]), and is useful in mathematical physics, being a natural spatially compact space-time which admits the action of the full conformal group.

Some of the results to be presented here are treated somewhat differently and in more specific situations in an excellent foundational work by Khinchin [Kh], who treats asymptotic distributions for photon systems in Minkowski space. The asymptotic spectrum of the Hamiltonian there is approximated to be concentrated on the integers; such a procedure suffices for consideration of energy observables for photons in three dimensional geometries, although it must be considered somewhat heuristic. It is however not adequate for treatment of more general situations (as will be seen here), since the very volatile dependence of distributions on spectral densities near 0 is not visible in such an analysis. In general terms, however, the present work extends many ideas pioneered by Khinchin.

The reading of any chapter is perhaps best done in two stages, the first involving only brief inspections of technical aspects of proofs, and the second involving a more thorough reading. Technicalities are often unavoidable in the proofs of theorems whose hypotheses are minimally technical.

We now provide a brief explanation of the structure of this monograph. The remainder of this chapter provides a mathematical-physical framework. This includes a description of Fock space, and a rigorous presentation of elementary results on quantum statistical probability distributions. For example, we verify the common assumption that occupation numbers are independent random variables. Some of these results may be found (in somewhat less thorough form) in statistical physics texts. Chapter Two presents some novel aspects of calculating distributions of observables (still in discrete situations), and presents fundamentals of the continuum limit. The basics of integration theory of random variable-valued functions are developed in Chapter 3. Chapters 4 and 5 give applications to calculating observable distributions, including criteria for normality and non-normality, and Chapter 6 provides physical applications. Here explicit distributions are calculated, and a rigorous Planck law is derived for Bose and Fermi ensembles. It is shown that the Planck laws are essentially independent of the intrinsic geometry of large systems. Chapters 7 and 8 develop an analogous Lebesgue theory of integration, and Chapter 9 provides further applications, in the more general framework.

We dispose of a few technical preliminaries. Throughout this work (except in §9.5) we make the physical assumption that the chemical potential of particles under consideration is 0; this does not involve essential loss of generality, as the general case can be treated similarly. Probability distributions are studied for particles obeying Bose and Fermi statistics. The two situations are similar, and are studied in parallel. The essentials of Segal's [Se3,4] formalism for free boson and fermion fields are used throughout and are described below.

We briefly mention some conventions. The symbols $\mathbf{Z}, \mathbf{N}, \mathbf{C}$, and $\mathbf{R}$ denote the integers and the natural, complex, and real numbers, respectively. The non-negative integers and reals are represented by $\mathbf{Z}^+$ and $\mathbf{R}^+$. The C^∞ functions with compact support are denoted by C_c^∞. The symbols $\mathcal{P}, \mathcal{E}$, and $\mathcal{V}$ denote probability measure, expectation and variance. The point mass at 0 is represented by δ_0, while $\Rightarrow$ and $[\cdot]$ denote convergence in law (or weak convergence) and the greatest integer function, respectively. The spectrum of an operator A is $\sigma(A)$; the words "positive" and "non-negative" are used interchangeably, as are "increasing" and "non-decreasing", etc. The abbreviations r.v., d.f., ch.f and a.s., mean "random variable", "distribution function", "characteristic function," and "almost surely".

The notation w-lim denotes a weak limit, i.e., limit in distribution. $N(a,b)$ denotes the normal distribution with mean a and variance b. If X is an r.v. or a distribution, then F_X denotes its d.f.

We use the notation $f(x) = O(x) \quad (x \to a)$ if $\frac{f(x)}{O(x)}$ is bounded as $x \to a$. We use $f(x) = o(x) \quad (x \to a)$ if $\frac{o(x)}{f(x)} \to 0$ as $x \to a$.

A question may arise as to the dimensions of physical quantities. The system of units will be entirely general, unless otherwise specified. For instance, the energy E and inverse temperature β may be interpreted in any units inverse to each other. Also, a preference for setting $c = 1$ will be evident in various places.

§1.2. The Free Boson and Fermion Fields Over a Hilbert Space

We now present relevant aspects of free boson and fermion fields in their particle representations. No proofs will be supplied; a more detailed and general description is given by Segal [Se3, Se4].

Let $\mathcal{H}$ be a separable complex Hilbert space; for notational convenience and without loss of generality assume $\mathcal{H}$ is infinite dimensional. For $n \in \mathbf{N}$ let

$$\mathcal{K}_n = \mathcal{H} \otimes \mathcal{H} \otimes \ldots \otimes \mathcal{H} = \bigotimes_{i=1}^{n} \mathcal{H} \tag{1.1}$$

be the n-fold tensor product of $\mathcal{H}$ with itself, and $V_n^{(B)}(\cdot)$ be the unitary representation of the symmetric group Σ_n of order n on $\mathcal{K}_n$ which is uniquely determined by the property

$$V_n^{(B)}(\sigma)\left(\bigotimes_{i=1}^{n} x_i\right) = \bigotimes_{i=1}^{n} x_{\sigma^{-1}(i)} \qquad (\sigma \in \Sigma_n;\ x_i \in \mathcal{H});$$

$V_n^{(B)}(\sigma)$ thus permutes tensors. The (closed) subspace $\mathcal{K}_n^{(B)}$ consisting of elements left invariant by $\left\{V_n^{(B)}(\sigma) : \sigma \in \Sigma_n\right\}$ is the *n-fold symmetrized tensor product* of $\mathcal{H}$ with itself. By convention $\mathcal{K}_0^{(B)} = \mathbf{C}$.

For $x_1, \ldots, x_n \in \mathcal{H}$.

$$x_1 \vee x_2 \vee \ldots \vee x_n \equiv \bigvee_{i=1}^{n} x_i \equiv \frac{1}{n!} \sum_{\sigma \in \Sigma_n} V_n^{(B)}(\sigma)\left(\bigotimes_{i=1}^{n} x_i\right) \tag{1.2}$$

is the orthogonal projection of $\bigotimes_{i=1}^{n} x_i$ into $\mathcal{K}_n^{(B)}$; the latter is clearly spanned by vectors of the form (1.2). Note that for $x_i \in \mathcal{H}$ and $\sigma \in \Sigma_n$,

$$\bigvee_{i=1}^{n} x_i = \bigvee_{i=1}^{n} x_{\sigma(i)}. \tag{1.3}$$

The direct sum of symmetrized tensor products over all orders

$$\mathcal{K}_B \equiv \bigoplus_{n=0}^{\infty} \mathcal{K}_n^{(B)}$$

is the *(Hilbert) space of symmetrized tensors* over $\mathcal{H}$.

Any unitary U on $\mathcal{H}$ can be lifted to a unitary $\Gamma_n^{(B)}(U) : \mathcal{K}_n^{(B)} \to \mathcal{K}_n^{(B)}$ which is defined uniquely by

$$\Gamma_n^{(B)}(U) \bigvee_{i=1}^{n} x_i = \bigvee_{i=1}^{n} (Ux_i) \qquad (x_i \in \mathcal{H}); \tag{1.4}$$

$\Gamma_n^{(B)}(U)$ is simply the restriction to $\mathcal{K}_n^{(B)}$ of the n-fold tensor product of U with itself; we append the convention that $\Gamma_0^{(S)}(U) : \mathbf{C} \to \mathbf{C}$ is the identity. We define

$$\Gamma_B(U) = \bigoplus_{n=0}^{\infty} \Gamma_n^{(B)}(U),$$

so that for $x_n \in \mathcal{K}_n$,

$$\Gamma_B(U)\left(\bigoplus_{n=0}^{\infty} x_n\right) = \bigoplus_{n=0}^{\infty} \Gamma_n^{(B)}(U) x_n.$$

A self-adjoint operator A in $\mathcal{H}$ is naturally lifted to $\mathcal{K}_B$ as the self-adjoint generator of the one parameter unitary group $\Gamma_B(e^{itA})$:

$$d\Gamma_B(A) \equiv \frac{1}{i}\frac{d}{dt}\Gamma_B(e^{itA})\bigg|_{t=0}, \tag{1.5}$$

with the derivative taken in the strong operator topology.

The definitions for antisymmetric statistics are fully analogous to those above. Let $s(\sigma)$ denote the sign of $\sigma \in \Sigma_n$, and $V_n^{(F)}(\cdot)$ be the unitary representation of Σ_n on $\mathcal{K}_n$ defined by

$$V_n^{(F)}(\sigma)\left(\bigotimes_{i=1}^{n} x_i\right) = s(\sigma)\bigotimes_{i=1}^{n} x_{\sigma^{-1}(i)} \qquad (\sigma \in \Sigma_n,\ x_i \in \mathcal{H}).$$

Let $\mathcal{K}_n^{(F)}$ denote the subspace of elements left invariant by $\left\{V_n^{(F)}(\sigma) : \sigma \in \Sigma_n\right\}$, with $\mathcal{K}_0^{(F)} \equiv \mathbf{C}$; the collection

$$x_1 \wedge x_2 \wedge \ldots \wedge x_n \equiv \bigwedge_{i=1}^{n} x_i \equiv \frac{1}{n!}\sum_{\sigma \in \Sigma_n} V_n^{(F)}(\sigma)\left(\bigotimes_{i=1}^{n} x_i\right) \qquad (x_i \in \mathcal{H}); \tag{1.6}$$

spans $\mathcal{K}_n^{(F)}$. The *space of antisymmetrized tensors* over $\mathcal{H}$ is

$$\mathcal{K}_F = \bigoplus_{n=0}^{\infty} \mathcal{K}_n^{(F)}. \tag{1.7}$$

If U is unitary on $\mathcal{H}$, then $\Gamma_n^{(F)}(U): \mathcal{K}_n^{(F)} \to \mathcal{K}_n^{(F)}$ is defined by

$$\Gamma_n^{(F)}(U)\left(\bigwedge_{i=1}^{n} x_i\right) = \bigwedge_{i=1}^{n} Ux_i,$$

with $\Gamma_0^{(F)}: \mathbf{C} \to \mathbf{C}$ the identity,

$$\Gamma_F(U) \equiv \bigoplus_{n=0}^{\infty} \Gamma_n^{(F)}(U)$$

is the lifting of U to $\mathcal{K}_n^{(F)}$. If A is self-adjoint in $\mathcal{H}$, then $d\Gamma_F(A)$ is the generator of the unitary group $\Gamma_F(e^{itA})$.

Henceforth statements not specifically referring to the (anti-)symmetric constructions will hold under both statistics; in particular this will hold when subscripts B and F are omitted.

DEFINITION 1.1: The operator $d\Gamma(A)$ is the *quantization* of A.

The map $d\Gamma$ acts linearly on bounded and unbounded self-adjoint operators to the extent that if $\{P_j\}_{j\in\mathbf{N}}$ are mutually orthogonal projections in $\mathcal{H}$ and $\{E_j\}_{j\in\mathbf{N}} \subset \mathbf{R}$, then

$$d\Gamma\left(\sum_{j=1}^{\infty} E_j P_j\right) = \sum_{j=1}^{\infty} E_j d\Gamma(P_j); \tag{1.8}$$

this fact will later prove useful. Physically, $(\mathcal{K}, d\Gamma(A))$ is the Hilbert space of states together with the Hamiltonian of a many-particle non-interacting system each particle of which has states in $\mathcal{H}$ and time evolution governed by A.

If $G = \{g_j\}_{j\in\mathbf{N}}$ is an orthonormal basis for $\mathcal{H}$ then orthonormal bases for $\mathcal{K}_n$ are given by

$$\begin{aligned} \mathcal{B}_n^{(B)} &= \left\{\bigvee_{i=1}^{n} g_{j_i} : j_1 \le j_2 \le \ldots \le j_n;\ j_i \in \mathbf{N}\right\} && \text{(symmetric)} \\ \mathcal{B}_n^{(F)} &= \left\{\bigwedge_{i=1}^{n} g_{j_i} : j_1 < j_2 < \ldots < j_n;\ j_i \in \mathbf{N}\right\} && \text{(antisymmetric)} \end{aligned} \tag{1.9}$$

The basis $\mathcal{B}_n^{(B)}$ can be represented through the correspondence

$$\bigvee_{i=1}^{n} g_{j_i} \leftrightarrow (m_1, m_2, \ldots) = \mathbf{n} \qquad (g_{j_i} \in G),$$

is multiplied out, it coincides with (1.13), so the latter converges if and only if

$$\sum_{j=0}^{\infty} e^{-\beta E_j} < \infty,$$

and no E_j are 0. The proof in the antisymmetric case is similar.

We thus have

$$K \equiv \operatorname{tr} e^{-\beta d\Gamma(A)} = \begin{cases} K_B = \prod_{j=1}^{\infty}(1 - e^{-\beta E_j})^{-1} & \text{(symmetric)} \\ K_F = \prod_{j=1}^{\infty}(1 + e^{-\beta E_j}) & \text{(antisymmetric)} \end{cases}. \tag{1.15}$$

COROLLARY 1.4.1: *If $e^{-\beta d\Gamma(A)}$ is trace class, then A has pure point spectrum and finite multiplicities.*

CHAPTER 2

VALUE FUNCTIONS ON A CANONICAL ENSEMBLE

Throughout this chapter A is a positive self-adjoint (energy) operator in $\mathcal{H}$, satisfying (*i*) or (*ii*) of Proposition 1.4. In particular A has pure point spectrum $\{E_j\}_{j=1}^{\infty}$, and an orthonormal eigenbasis $G = \{g_j\}_{j \in \mathrm{N}}$, with $Ag_j = E_j g_j$. Assume $\mathcal{H}$ is separable and (for notational convenience) infinite dimensional, and define the number operator $N_j = d\Gamma(P_j)$ as before. The basis $\mathcal{N}$ of $\mathcal{K}$ corresponding to G is given in (1.10); its generic elements will be $\mathbf{n} = (m_1, m_2, \ldots)$. Define

$$\rho = \frac{e^{-\beta d\Gamma(A)}}{\operatorname{tr} e^{-\beta d\Gamma(A)}}. \tag{2.1}$$

If a is the value function of $d\Gamma(A)$, and n_j that of N_j, then

$$h(a) = \sum_{j=1}^{\infty} h(E_j) n_j. \tag{2.2}$$

More generally, if C commutes with A and $Cg_j = c_j g_j$, $d\Gamma(C)$ has value function $\sum_{j=1}^{\infty} c_j n_j$.

DEFINITION 2.1: The r.v. n_j is the jth *occupation number* in the canonical ensemble over A.

Note that in the ensemble, according to (1.12),

$$\mathcal{P}(\mathbf{n}) = \langle \rho\mathbf{n}, \mathbf{n} \rangle = \frac{1}{\operatorname{tr} e^{-\beta d\Gamma(A)}} \langle e^{-\beta \sum_{j=1}^{\infty} E_j m_j} \mathbf{n}, \mathbf{n} \rangle = \frac{1}{K} e^{-\beta \sum E_j m_j}. \tag{2.3}$$

§2.1. Physical Interpretation of Value Functions

The canonical ensemble over A at inverse temperature $\beta > 0$ describes a non-self-interacting system of particles in equilibrium at temperature $T = \frac{1}{\beta k}$ (k is Boltzmann's constant) with particles whose individual time evolution is governed by the Hamiltonian A. If the operator C, representing a physical observable for single particles, commutes with A, $d\Gamma(C)$ is the observable representing "total amount" of C. Precisely,

$$d\Gamma(C)\mathbf{n} = \left(\sum_{j=1}^{\infty} m_i \langle Cg_i, g_i \rangle \right) \mathbf{n}, \tag{2.4}$$

$\langle Cg_j, g_j \rangle$ representing the value of C in the state g_j,

The value function of $d\Gamma(C)$ is a random variable on the canonical ensemble, representing the probabilistic nature of $d\Gamma(C)$. By its definition

$$c(\mathbf{n}) = \sum_{j=1}^{\infty} m_i \langle C g_j, g_j \rangle. \tag{2.5}$$

In particular, the occupation number n_j corresponding to g_j is the r.v. on $(\mathcal{N}, \mathcal{P})$ which on $\mathbf{n}$ takes the value m_j, i.e., the number of particles in state g_j. The value of $\Delta_j = E_j n_j$ on $\mathbf{n}$ is $E_j m_j$, and represents the "total energy" of the particles in state g_j.

§2.2. Distribution of Occupation Numbers

If n_i denote occupation numbers and if $\mathbf{t} = (t_1, t_2, \ldots)$, $\bar{\mathbf{n}} = (n_1, n_2, \ldots)$ and $\mathbf{t} \cdot \bar{\mathbf{n}} = \sum t_i n_i$, then according to (2.3) the joint characteristic function of $\{n_j\}_{j \in \mathrm{N}}$ is

$$\Phi(\mathbf{t}) = \mathcal{E}(e^{i\mathbf{t}\cdot\bar{\mathbf{n}}}) = \frac{1}{K} \sum_{\mathbf{n} \in \mathcal{N}} e^{i\mathbf{t}\cdot\mathbf{n} - \beta \mathbf{E}\cdot\mathbf{n}}, \tag{2.6}$$

where $\mathbf{E} = (E_1, E_2, \ldots)$, and $\mathcal{E}$ denotes mathematical expectation. The right side factors to give

$$\Phi(\mathbf{t}) = \frac{1}{K} \prod_{j=1}^{\infty} \begin{cases} \left(1 - e^{-\beta E_j + i t_j}\right)^{-1} & \text{(symmetric)} \\ \left(1 + e^{-\beta E_j + i t_j}\right) & \text{(antisymmetric)} \end{cases}. \tag{2.7}$$

Thus, occupation numbers n_j are independent r.v.'s with characteristic functions

$$\Phi_j(t) = \begin{cases} \left(1 - e^{-\beta E_j}\right)\left(1 - e^{-\beta E_j + it}\right)^{-1} \\ \left(1 + e^{-\beta E_j}\right)^{-1}\left(1 + e^{(-\beta E_j + it)}\right) \end{cases}. \tag{2.8}$$

By inversion of (2.8), n_j is geometric in the first case, with

$$\mathcal{P}(n_j = l) = \left(1 - e^{-\beta E_j}\right) e^{-\beta E_j l} \qquad (l = 0, 1, \ldots), \tag{2.9}$$

$$\mathcal{E}(n_j) = \frac{1}{e^{\beta E_j} - 1}, \qquad \mathcal{V}(n_j) = \frac{e^{\beta E_j}}{\left(e^{\beta E_j} - 1\right)^2}, \tag{2.10}$$

where $\mathcal{V}$ denotes variance. Under antisymmetric statistics n_j is Bernoulli, with

$$\mathcal{P}(n_j = l) = \begin{cases} \left(e^{\beta E_j} + 1\right)^{-1}; & l = 1 \\ \left(1 + e^{-\beta E_j}\right)^{-1}; & l = 0 \end{cases} \tag{2.11}$$

$$\mathcal{E}(n_j) = \frac{1}{e^{\beta E_j} + 1}, \qquad \mathcal{V}(n_j) = \frac{e^{\beta E_j}}{\left(e^{\beta E_j} + 1\right)^2}.$$

The above distributions (specifically the independence of the occupation numbers) explicitly verify facts which have been used in the physical literature for some time.

Rewriting of (2.3) shows that if

$$\mathcal{N}_l = \left\{ \mathbf{n} = (m_1, m_2, \ldots) : |\mathbf{n}| \equiv \sum_{j=1}^{\infty} m_j = l \right\}, \tag{2.12}$$

the particle number value function $n = \sum_{j=1}^{\infty} n_j$ satisfies

$$\mathcal{P}(n = l) = \frac{1}{K} \sum_{\mathbf{n} \in \mathcal{N}_l} e^{-\beta \mathbf{E} \cdot \mathbf{n}} = \frac{1}{K} \sum_{1 \leq j_1 \leq \ldots \leq j_l} e^{-\beta \sum_{m=1}^{l} E_{j_m}} \quad (l \in \mathbf{Z}^+). \tag{2.13}$$

under symmetric statistics, with $\leq$ replaced by $<$ in the antisymmetric case. There is a method of expressing (2.13) more simply for small values of l which, however, becomes more complicated as l increases, and is not discussed here.

The total energy

$$\Delta = \mathbf{E} \cdot \overline{\mathbf{n}} = \sum E_j n_j$$

is a discrete r.v. with

$$P(\Delta = b) = \frac{1}{K} \sum_{\mathbf{n} \in \mathcal{N}_b} e^{-\beta b} = \frac{e^{-\beta b} d(b)}{K},$$

where $d(b)$ is the cardinality of $\mathcal{N}_b = \{\mathbf{n} \in \mathcal{N} : \mathbf{n} \cdot \mathbf{E} = b\}$.

When A has uniformly spaced spectrum $E_j = \epsilon j$, (physically interesting in one dimensional situations), these specialize in the symmetric case to

$$\begin{aligned}
\mathcal{P}(n = l) &= \frac{1}{K_B} \sum_{1 \leq j_1 \ldots \leq j_l} e^{-\beta\epsilon \sum_{m=1}^{l} j_m} \\
&= \frac{1}{K_B} \sum_{1 \leq j_1 \ldots \leq j_{l-1}} \left(e^{-\beta\epsilon \sum_{m=1}^{l-1} j_m} \sum_{j_l \geq j_{l-1}} e^{-\beta\epsilon j_l} \right) \\
&= \frac{1}{K_B} \sum_{1 \leq j_1 \leq \ldots \leq j_{l-1}} e^{-\beta\epsilon \sum_{m=1}^{l-1} j_m} \frac{e^{-\beta\epsilon j_{l-1}}}{1 - e^{-\beta\epsilon}} \\
&= \frac{1}{K_B} \sum_{1 \leq j_1 \leq \ldots \leq j_{l-2}} e^{-\beta\epsilon \sum_{m=1}^{l-2} j_m} \frac{e^{-2\beta\epsilon j_{l-2}}}{(1 - e^{-\beta\epsilon})(1 - e^{-2\beta\epsilon})} \\
&= \cdots = \frac{1}{K_B} e^{-l\beta\epsilon} \prod_{j=1}^{l} \left(1 - e^{-j\beta\epsilon}\right)^{-1} = e^{-l\beta\epsilon} \prod_{j=l+1}^{\infty} (1 - e^{-j\beta\epsilon}).
\end{aligned} \tag{2.14a}$$

Similarly in the antisymmetric case

$$\mathcal{P}(n = l) = \frac{1}{K_F} e^{\frac{-l(l+1)}{2}} \prod_{j=1}^{l} (1 - e^{-\beta \epsilon j}). \tag{2.14b}$$

Note that the tail of the number r.v. in (2.14) is geometric under symmetric statistics, and resembles a discretized Gaussian in the antisymmetric case.

It is easily seen in this case that

$$\mathcal{P}(\Delta = \epsilon l) = \begin{cases} \frac{e^{-\beta \epsilon l} p_l}{K_B} & \text{(symmetric)} \\ \frac{e^{-\beta \epsilon l} q_l}{K_F} & \text{(antisymmetric)} \end{cases} \tag{2.15}$$

where p_l (q_l) is the number of ways of representing l as a sum of (distinct) positive integers. The asymptotics of the combinatorial functions p_l and q_l are tabulated in books of mathematical functions (see [AS]).

§2.3. Formalism of Asymptotic Spectral Densities

In many situations in which a self-adjoint operator has large or asymptotically infinite spectral density, use of a spectral measure on the real line is necessary for interpretation of sums indexed by the spectrum, or mediating more detailed spectral information. A classical example occurs in Sturm-Liouville systems on $[0, \infty)$, whose "natural" spectral measures for eigenfunction expansions are defined in terms of limits of spectral densities of their restrictions to intervals (see [GS]). In the study of sums such as $n = \sum n_j$, a similar situation occurs if the spectrum $\{E_j\}$ of A is very dense or continuous, e.g., if A is an elliptic differential operator on a "large" Riemannian manifold. See Simon's review article [Si, Sections C6, C7] for a perspective on this problem in Schrödinger theory. Situations such as these certainly predominate in macroscopic systems. In such cases, the appropriate probabilistic context for study of probability distributions in canonical ensembles involves integrals (rather than sums) of independent random variables with respect to appropriate spectral measures.

We begin with preliminary notions and facts. Henceforth $[\cdot]$ will denote the greatest integer function.

DEFINITION 2.2: A *spectral measure* μ is a σ-finite measure on $\mathbf{R}^+ = [0, \infty]$; $F(E) = \mu[0, E]$ is the *spectral function* of μ. The operator A_ϵ with point spectrum consisting of the discontinuities of $\left[\frac{F(E)}{\epsilon}\right]$, each discontinuity E having multiplicity $\left[\frac{F(E)}{\epsilon}\right] - \left[\frac{F(E^-)}{\epsilon}\right]$, is the *$\epsilon$-discrete operator* corresponding to μ.

The operator A_ϵ has eigenvalue density essentially given by $\frac{\mu}{\epsilon}$.

The correspondence between spectral measures μ and nets $\{A_\epsilon\}_{\epsilon>0}$ is bijective. Indeed, the ϵ-net of functions $\epsilon\left[\frac{F(E)}{\epsilon}\right]$ converges uniformly to F, determining μ.

We now investigate when A_ϵ defines a canonical ensemble, using criteria established in Chapter 1. We require the following convention: the domain of integration in a Stieltjes integral $\int_a^b g(x)df(x)$ includes endpoints, and

$$\int_{a^-}^{b} g(x)df(x) \equiv \lim_{\epsilon\to 0}\int_{a-\epsilon}^{b} g(x)df(x). \tag{2.16}$$

We begin with a technical lemma.

LEMMA 2.3 *Let $f(\cdot)$ and $g(\cdot)$ be defined on $[a,b]$, $g(\cdot)$ be monotone non-increasing and non-negative, $f(\cdot)$ be monotone and $g(\cdot)$ be Stieltjes integrable with respect to $f(\cdot)$ and $[f(\cdot)]$. Then*

$$\left|\int_a^b g(x)df(x) - \int_a^b g(x)d[f(x)]\right| \le g(a). \tag{2.17}$$

This holds if "a" is replaced by "a^-", "b" by "b^-" or if $[a,b]$ is replaced by $[a,\infty)$.

Proof: Assume without loss that $f(\cdot)$ is non-decreasing, and that $f(\cdot)$ is continuous on $[a,b]$; some inessential complications occur if the latter fails. Assume $b=\infty$ (otherwise f and g can be appropriately extended), and let

$$a \le x_1 < x_2 < x_3 < \ldots, \tag{2.18}$$

where $\{x_k\}$ are the discontinuity points of $[f(\cdot)]$. If $\{x_k\}$ is empty the assertion is trivial. Otherwise let

$$S_1 = [a,x_1],\ S_2 = (x_1,x_2],\ S_3 = (x_2,x_3],\ldots .$$

Then

$$\int_a^\infty g(x)df(x) = \sum_{k=1}\int_{S_k} g(x)df(x); \qquad \int_a^\infty g(x)d[f(x)] = \sum_{k=1} g(x_k), \tag{2.19}$$

and

$$g(x_k) \ge \int_{S_{k+1}} g(x)df(x) \ge g(x_{k+1}) \qquad (k=1,2,\ldots). \tag{2.20}$$

Hence

$$\left|\sum_{k=1} g(x_k) - \sum_{k=2}\int_{S_k} g(x)df(x)\right| \le g(a). \tag{2.21}$$

The fact that

$$\int_{S_1} g(x)df(x) \le g(a) \tag{2.22}$$

and the first inequality in (2.20) imply (2.18); the remaining assertions follow similarly.∎

COROLLARY 2.4: *If instead of monotone g is of bounded variation, then*

$$\left| \int_a^b g(x) df(x) - \int_a^b g(x) d[f(x)] \right| \leq 2g(a).$$

THEOREM 2.5: *Let μ be a spectral measure and A_ϵ the corresponding ϵ-discrete operator. Let $f(\cdot)$ be a non-increasing function. Then $f(A_\epsilon)$ is trace class if and only if*

$$\int_0^\infty f(E) d\mu(E) < \infty. \tag{2.23}$$

Proof: If $f(A_\epsilon)$ is trace class, then

$$\mathrm{tr}\, f(A_\epsilon) = \int_{0^-}^\infty f(E) d\left[\frac{F(E)}{\epsilon}\right],$$

while

$$\int_0^\infty f(E) d\mu(E) = \epsilon \int_{0^-}^\infty f(E) d\left(\frac{F(E)}{\epsilon}\right). \tag{2.24}$$

If $F(\cdot)$ is bounded on $[0, \infty)$, the last expression is finite. Otherwise, the lemma implies

$$\epsilon \int_{0^-}^\infty f(E) d\left[\frac{F(E)}{\epsilon}\right] \leq \epsilon \int_{0^-}^\infty f(E) d\left[\frac{F(E)}{\epsilon}\right] + f(0^-) = \epsilon\, (\mathrm{tr}\, f(A_\epsilon)) + f(0^-) < \infty.$$

Conversely if (2.23) holds, then

$$\mathrm{tr} f(A_\epsilon) = \int_{0^-}^\infty f(E) d\left[\frac{F(E)}{\epsilon}\right] \leq \int_{0^-}^\infty f(E) d\left(\frac{F(E)}{\epsilon}\right) + f(0^-) < \infty. \blacksquare \tag{2.25}$$

COROLLARY 2.6: *The operator $f(A_\epsilon)$ is trace class for some $\epsilon > 0$ if and only if it is for all $\epsilon > 0$.*

COROLLARY 2.7: *The operator $e^{-\beta A_\epsilon}$ is trace class if and only if*

$$\int_0^\infty e^{-\beta E} d\mu(E) < \infty.$$

DEFINITIONS 2.8: A spectral measure μ is 0-*free* if $\mu(\{0\}) = 0$.

Note that μ is 0-free if and only if A_ϵ is 0-free for all ϵ. This fact together with Corollary 2.7 and Proposition 1.4 completely characterizes those spectral measures which define canonical ensembles.

CHAPTER 3

INTEGRALS OF INDEPENDENT RANDOM VARIABLES.

In the limit of continuous spectrum for a Hamiltonian A, sums of r.v.'s indexed by the spectrum, such as value functions of quantizations of observables commuting with A, are most profitably represented as integrals for both notational and conceptual reasons. We now develop theory and properties of these to the extent that they are useful later. These results will be in a more natural probabilistic setting in Chapters 7 and 8, where integrals of r.v.-valued functions over a measure space will be studied.

§3.1. General Theory of One Parameter Families.

For $E \geq 0$, let $X(E)$ be a one-parameter family of independent r.v.'s and μ a spectral measure on $\mathbf{R}^+$. We integrate X with respect to μ as follows. For each $\epsilon > 0$ let $\mathcal{F}_\epsilon = \{E_{\epsilon j}\}_{j\in J_\epsilon}$ be positive numbers (not necessarily distinct) such that the cardinality $N_\epsilon(a,b)$ of $\mathcal{F}_\epsilon \cap \{[a,b]\}$ satisfies

$$\epsilon N_\epsilon(a,b) \underset{\epsilon\to 0}{\to} \mu[a,b] \tag{3.1a}$$

for all $0 \leq a < b$ (b may be ∞). To avoid pathologies, assume also the existence of numbers M, $k > 0$ such that for any interval I whose measure exceeds M,

$$\epsilon N_\epsilon(I) \leq k\mu(I). \tag{3.1b}$$

DEFINITION 3.1: $\{\mathcal{F}_\epsilon\}_{\epsilon > 0}$ *is a spectral* μ-net.

Given a function $\phi(\epsilon)$ on positive ϵ, we define the probability distribution

$$\int_{\mathbf{R}^+} X(E)\phi(d\mu) \equiv \text{w-}\lim_{\epsilon\to 0} \phi(\epsilon)\sum_j X(E_{\epsilon j})$$

with sums on the right defined only if order-independent. The limit is in the topology of convergence in law, with the integral defined only if independent of $\mathcal{F}_\epsilon$.

If f is real-valued on $\mathbf{R}^+$, the Riemann-Stieltjes integral of f with respect to μ can be defined as

$$R\int_{\mathbf{R}^+} f(E)d\mu \equiv \lim_{\mu(P)\to 0} \sum f(E_i)\mu(\Delta E_i) \tag{3.2}$$

with P a partition of $\mathbf{R}^+$ into intervals ΔE_i, $E_i \in \Delta E_i$, $\mu(P) = \sup_i \mu(\Delta E_i)$, and the limit is defined only when independent of P and E_i, as well as order of summation (i.e., convergence is absolute). In evaluating the right side of (3.2) and all similar sums we apply the convention: if μ has an atom at $E \geq 0$, a partition P may formally divide E into degenerate intervals $\{\Delta E_i\}_{i\epsilon J}$, each consisting of just the point E, such that $\sum_J \mu(\Delta E_i) = \mu(\{E\})$; the union of such an interval with an adjacent non-degenerate interval is also allowable. P is an *even partition* if $\mu(E_i) = \mu(E_j)$, except possibly when E_j is a rightmost non-empty interval, which is allowed smaller measure.

THEOREM 3.2: *If $X(E)$ is a real number (i.e., a point mass) for each E, then $\int X(E)d\mu$ coincides exactly with ${}_R\int X(E)d\mu$.*

Proof: Assume that ${}_R\int X(E)d\mu$ exists, and let $\mathcal{F}_\epsilon = \{E_{\epsilon j}\}_{j\epsilon J_\epsilon}$ be a spectral μ-net. For $n \in \mathbf{N}$ let $P_n = \{\Delta E_{ni}\}_{i\epsilon I_n}$ be a sequence of even partitions of $\mathbf{R}^+$ following the above conventions, such that $\mu(P_n) \underset{n\to\infty}{\to} 0$. The indexing set $I_n \subset \mathbf{N}$ may be finite or the whole of $\mathbf{N}$. For $i \in I_n$, let $\tilde{E}_{ni}$, $E^*_{ni} \in \Delta E_{ni}$ satisfy

$$X(E^*_{ni}) - \frac{1}{ni^2} \leq X(E) \leq X(\tilde{E}_{ni}) + \frac{1}{ni^2} \qquad (E \in \Delta E_{ni}), \tag{3.3}$$

and define

$$a_n = \sum_{i\in I_n} \frac{\mu(\Delta E_{ni})}{ni^2}$$

$$A_n = \sum_{i\epsilon I_n} X(E^*_{ni})\mu(\Delta E_{ni}), \qquad B_n = \sum_{i\epsilon I_n} X(\tilde{E}_{ni})\mu(\Delta E_{ni}).$$

If μ has compact support then I_n is finite, and by (3.1) and (3.3), the distance between the number $\epsilon\sum_{j\in J_\epsilon} X(E_{\epsilon j})$ and the interval $[A_n - a_n, B_n + a_n]$ approaches 0 as $\epsilon \to 0$. Since $a_n \underset{n\to\infty}{\to} 0$,

$$[A_n + a_n, B_n - a_n] \underset{n\to\infty}{\to} {}_R\int X(E)d\mu,$$

and the result follows here, since the μ-net $\mathcal{F}_\epsilon$ was arbitrary. For the general case, we must show

$$\epsilon \sum_{|E_{\epsilon j}|>d} |X(E_{\epsilon j})| \underset{d\to\infty}{\to} 0,$$

uniformly in ϵ. Let $P = \{\Delta E_j\}$ be an even partition of $\mathbf{R}^+$ such that

$$\sum_j |X(E_j)|\mu(\Delta E_j) < C \tag{3.4}$$

for any set of $E_j \in \Delta E_j$. By (3.1*b*)

$$\epsilon N_\epsilon(\Delta E_j) \leq k \max(M, \mu(\Delta E_j)),$$

so that

$$\epsilon \sum_{E_{\epsilon j} \in \Delta E_i} |X(E_{\epsilon j})| \leq k \max(M, \mu(\Delta E_j)) \sup_{E \in \Delta E_i} |X(E)|.$$

Since P is even and by (3.4),

$$\epsilon \sum_{|E_{\epsilon j}| > d} |X(E_{\epsilon j})| \leq k \sum_{\Delta E_j \cap [d,\infty] \neq \phi} \sup_{E \in \Delta E_j} |X(E)| \max(M, \mu(\Delta E_j)) \underset{d \to \infty}{\to} 0, \tag{3.5}$$

convergence being manifestly uniform in ϵ.

Conversely, assume $\int X(E) d\mu$ exists. Let $P_n = \{\Delta E_{nj}\}$ be a sequence of partitions of $\mathbf{R}^+$ with $\mu(P_n) \underset{n \to \infty}{\to} 0$, and let $E_{nj} \in \Delta E_{nj}$. There exists an even partition $P^* = \{\Delta E_j^*\}$ such that

$$Q = \sum_j \sup_{E \in \Delta E_j^*} |X(E)| \mu(\Delta E_j^*) < \infty;$$

hence if $\mu(P_n) < \mu(P^*)$, then

$$\sum_j \sup_{E \in \Delta E^*_{nj}} |X(E)| \mu(\Delta E_{ni}) \leq 3Q. \tag{3.6}$$

Let $\overline{\mathcal{F}}_\epsilon = \{\overline{E}_{\epsilon i}\}$ be chosen such that for each $\overline{E}_{\epsilon i}$ there is an integer n_ϵ (depending only on ϵ) and a j such that

a) The cardinality of $\{i : \overline{E}_{\epsilon i} = E_{n_\epsilon j}\}$ is $\left[\frac{\mu(\Delta E_{n_\epsilon j})}{\epsilon}\right]$ where $[\cdot]$ denotes the greatest integer function, and for each i, $E_{\epsilon i} = E_{n_\epsilon j}$ for some j.

b) If $J_\epsilon = \{j : \mu(\Delta E_{n_\epsilon j}) < \epsilon n_\epsilon\}$ then $\sum_{j \epsilon J_\epsilon} |X(E_{n_\epsilon j})| \mu(\Delta E_{n_\epsilon j}) < \frac{1}{n_\epsilon}$.

c) $n_\epsilon \underset{\epsilon \to 0}{\to} \infty$ monotonically.

Then if J_ϵ^c is the complement of J_ϵ,

$$\begin{aligned} \left| \epsilon \sum_i X(\overline{E}_{\epsilon i}) - \sum_j X(E_{n_\epsilon j}) \mu(\Delta E_{n_\epsilon j}) \right| &= \left| \sum_j X(E_{n_\epsilon j}) \left\{ \epsilon \left[\frac{\mu(\Delta E_{n_\epsilon j})}{\epsilon} \right] - \mu(\Delta E_{n_\epsilon j}) \right\} \right| \\ &\leq \sum_{j \in J_\epsilon} |X(E_{n_\epsilon j})| \mu(\Delta E_{n_\epsilon j}) + \sum_{j \in J_\epsilon^c} |X(E_{n_\epsilon j})| \epsilon \\ &\leq \frac{1}{n_\epsilon} + \sum_{j \in J_\epsilon^c} |X(E_{\epsilon j})| \frac{\mu(\Delta E_{n_\epsilon j})}{n_\epsilon} \\ &\leq \frac{1}{n_\epsilon} \{1 + 3Q\} \underset{\epsilon \to 0}{\to} 0, \end{aligned} \tag{3.7}$$

the above holding for ϵ sufficiently small. Thus ${}_R\int X(E) d\mu$ exists and equals $\int X(E) d\mu$, completing the proof. ∎

As is clear from the standard central limit theorem, expectation of r.v.'s does not scale under independent sums in the same way as some other "linear" quantities, such as standard deviation. For this reason expectation will in general diverge in the formation of integrals (in which $\phi(\epsilon)$ may be non-linear), while all other linear parameters converge to define a limiting distribution. To avoid this inessential complication, we assume henceforth that integrals involve 0-mean random variables.

DEFINITION 3.3: If Y is an r.v. or probability distribution, Y^{*n} denotes the convolution of Y with itself n times.

Recall that a distribution function F is *stable* if for every b_1, b_2 and $a_1, a_2 > 0$, there exist constants b and $a > 0$ such that

$$F(a_1x + b_1) * F(a_2x + b_2) = F(ax + b). \tag{3.8}$$

Note that stability implies infinite divisibility.

PROPOSITION 3.4: *If* $Y = \int X(E)\phi(d\mu)$ *exists and is non-zero, then* Y *is stable,*

$$l_n = \lim_{\epsilon \to 0} \frac{\phi(\epsilon)}{\phi(n\epsilon)} \qquad (n = 1, 2, 3, \ldots)$$

exists, and $l_n Y^{*n} = Y$.

Proof: Fix n and let $\{\epsilon_k\}$ be a sequence such that $\epsilon_{k+1} < \frac{\epsilon_k}{n}$. Consider a spectral μ-net $\mathcal{F}_\epsilon = \{E_{\epsilon,j}\}_{j \in J_\epsilon}$ such that $\mathcal{F}_{\epsilon_k}$ consists of $\mathcal{F}_{\frac{\epsilon_k}{n}}$ with each element repeated n times. For $k = 1, 2, 3\ldots$, let

$$Y_{2k-1} = \phi(n\epsilon_k) \sum_{j \in J_{n\epsilon_k}} X(E_{n\epsilon_k, j})$$

$$Y_{2k} = \phi(\epsilon_k) \sum_{j \in J_{\epsilon_k}} X(E_{\epsilon_k, j}),$$

where $n\epsilon_k$ indicates a product.
Then

$$Y_{2k} = Y_{2k-1}^{*n} \frac{\phi(\epsilon_k)}{\phi(n\epsilon_k)}.$$

Letting $k \to \infty$, we conclude that

$$l_n = \lim_{k \to \infty} \frac{\phi(\epsilon_k)}{\phi(n\epsilon_k)}$$

exists, and that $l_n Y^{*n} = Y$. ∎

COROLLARY 3.5: *If Y of Proposition 3.4 has finite second moment then it is normal.*

Proof: The normals are the only stable distributions with finite second moment (see [GK]).∎

The following theorem generalizes Liapounov's theorem (see, e.g. [Ch]) to infinite sums of random variables, and will be instrumental in subsequent limit theorems. Recall that convergence in law of a net $\{X_i\}_{i\in I}$ of random variables or distributions is denoted by $X_i \Rightarrow M$ $N(m,\sigma^2)$ is the normal distribution with mean m and variance σ^2.

THEOREM 3.6: *For each $\epsilon > 0$, let $\{X_{\epsilon j}\}_{1\leq j\leq K_\epsilon}$ be independent zero-mean r.v.'s on the same probability space and $K_\epsilon \underset{\epsilon\to 0}{\to} \infty$ (or $K_\epsilon = \infty$). Let*

$$\gamma_{\epsilon j}^3 = \mathcal{E}\left(|X_{\epsilon j}|^3\right), \quad \sigma_{\epsilon j}^2 = \mathcal{E}\left(X_{\epsilon j}^2\right),$$

$$\gamma_\epsilon^3 = \sum_{j\in K_\epsilon} \gamma_{\epsilon j}^3, \quad \sigma_\epsilon^2 = \sum_{j=K_\epsilon} \sigma_{\epsilon j}^2.$$

Then if $\frac{\gamma_\epsilon}{\sigma_\epsilon} \underset{\epsilon\to 0}{\to} 0$, the sum $\sum_{j\in J_\epsilon} X_{\epsilon j}$ is asymptotically normal:

$$\frac{1}{\sigma_\epsilon}\sum_j X_{\epsilon j} \underset{\epsilon\to 0}{\Rightarrow} N(0,1). \tag{3.9}$$

Proof: Let $k_\epsilon \leq K_\epsilon$ be an integer such that (i)

$$\frac{\sum_{j=1}^{k_\epsilon}\gamma_{\epsilon j}^3}{\left(\sum_{j=1}^{k_\epsilon}\sigma_{\epsilon j}^2\right)^{\frac{3}{2}}} \underset{\epsilon\to 0}{\to} 0$$

and (ii)

$$\frac{\sum_{j=k_\epsilon+1}^{K_\epsilon}(\sigma_{\epsilon j}^2)}{\sum_{j=1}^{k_\epsilon}\sigma_{\epsilon j}^2} \underset{\epsilon\to 0}{\to} 0. \tag{3.10}$$

The sum $\sum_{j=1}^{k_\epsilon} X_{\epsilon j}$ is asymptotically normal by (i) and Liapounov's theorem:

$$\frac{\sum_{j=1}^{k_\epsilon} X_{\epsilon j}}{\sum_{j=1}^{k_\epsilon}(\sigma_{\epsilon j}^2)^{\frac{1}{2}}} \underset{\epsilon\to 0}{\Rightarrow} N(0,1).$$

By (ii) the same holds for

$$T_\epsilon \equiv \frac{\sum_{j=1}^{k_\epsilon} X_{\epsilon j}}{\left(\sum_{j=1}^{K_\epsilon}\sigma_{\epsilon j}^2\right)^{\frac{1}{2}}}. \tag{3.11}$$

The remainder

$$U_\epsilon \equiv \frac{\sum_{j=k_\epsilon+1}^{K_\epsilon} X_{\epsilon j}}{\left(\sum_{j=1}^{K_\epsilon} \sigma_{\epsilon j}^2\right)^{\frac{1}{2}}}. \tag{3.12}$$

satisfies $\mathcal{E}(U_\epsilon^2) \underset{\epsilon\to 0}{\to} 0$. Hence

$$Q_\epsilon \equiv T_\epsilon + U_\epsilon \underset{\epsilon\to 0}{\Rightarrow} N(0,1).\blacksquare \tag{3.13}$$

THEOREM 3.7: *If* ${}_R\int_0^\infty \mathcal{E}(X^2)d\mu$ *and* ${}_R\int_0^\infty \mathcal{E}\left(|X|^3\right)d\mu < \infty$, *then*

$$\int_0^\infty X(E)(d\mu)^{\frac{1}{2}} = N(0,v),$$

where $v = {}_R\int_0^\infty \mathcal{E}(X^2)d\mu$.

Proof: Let $\mathcal{F}_\epsilon = \{E_{\epsilon j}\}_{j\in J}$ for $\epsilon > 0$ be a spectral μ-net, and $X_{\epsilon j} \equiv X(E_{\epsilon j})$. Let $\sigma_{\epsilon j}$ and $\gamma_{\epsilon j}$ be as above. Then

$$\frac{\gamma_\epsilon^3}{\sigma_\epsilon^3} = \frac{\sum_j \mathcal{E}\left(|X_{\epsilon j}|^3\right)}{\left(\sum_j \sigma_{\epsilon j}^2\right)^{\frac{3}{2}}} = \epsilon^{\frac{1}{2}} \frac{\epsilon\sum_j \mathcal{E}\left(|X_{\epsilon j}|^3\right)}{\left(\epsilon\sum_j \mathcal{E}(X_{\epsilon j}^2)\right)^{\frac{3}{2}}} \underset{\epsilon\to 0}{\to} 0\cdot \frac{\int \mathcal{E}\left(|X|^3\right)d\mu}{\left(\int \mathcal{E}\left(|X|^2\right)d\mu\right)^{\frac{3}{2}}} = 0. \tag{3.14}$$

Hence

$$\int X(E)(d\mu)^{\frac{1}{2}} = \lim_{\epsilon\to 0} \epsilon^{\frac{1}{2}} \sum_j X_{\epsilon j} = \lim_{\epsilon\to 0} \frac{\epsilon^{\frac{1}{2}}\sum_j X_{\epsilon j}}{\left(\epsilon\sum_j \sigma_{\epsilon j}^2\right)^{\frac{1}{2}}} \cdot v^{\frac{1}{2}} = v^{\frac{1}{2}} N(0,1) = N(0,v);$$

the second equality follows by Theorem 3.2, and the third by Theorem 3.6 .∎

§ 3.2. Background for Singular Integrals.

Now we prove a result central to the singular integral theory in Chapters 4 and 5. A singular integral $\int X(E)\phi(d\mu)$ is one where $\mathcal{E}(X^2(E))$ is singular. Since such integrals fail to exist in general under the current definition, we make one which is more useful.

DEFINITION 3.8: Let $X(E)$ be a one parameter family of independent r.v.'s on $\mathbf{R}^+$, and μ a σ-finite measure on $\mathbf{R}^+$. Let $F(E) = \mu([0,E])$, and for each $\epsilon > 0$, let $\mathcal{F}_\epsilon = \{E_{\epsilon j}\}_{j=1}^\infty$ be the ordered set of discontinuities of $\left[\frac{F(E)}{\epsilon}\right]$, a jump discontinuity of size n being listed n times. We define

$$S\int_{\mathbf{R}^+} X(E)\phi(d\mu) \equiv \lim_{\epsilon\to 0} \phi(\epsilon) \sum_{j=1}^\infty X(E_{\epsilon j}). \tag{3.15}$$

THEOREM 3.9: *Suppose $\phi(\epsilon)$ and $\psi(\epsilon)$ are functions and*

$$\frac{\psi^3}{\phi} \underset{\epsilon\to 0}{\to} 0, \quad s\int \mathcal{E}\Big(|X|^3\Big)\phi(d\mu) < \infty, \quad 0 < s\int \mathcal{E}(X^2)\psi^2(d\mu) \equiv v < \infty.$$

Then $s\int X\psi(d\mu) = N(0, v)$.

Proof: By the hypotheses,

$$\frac{\phi(\epsilon)\sum_j \mathcal{E}\Big(|X_{\epsilon j}|^3\Big)}{\psi^3(\epsilon)\Big(\sum_j \mathcal{E}(X_{\epsilon j}^2)\Big)^{\frac{3}{2}}} \underset{\epsilon\to 0}{\to} K < \infty$$

Hence,

$$\frac{\sum_j \mathcal{E}\Big(|X_{\epsilon j}|^3\Big)}{\Big(\sum_j \mathcal{E}(X_{\epsilon j}^2)\Big)^{\frac{3}{2}}} \underset{\epsilon\to 0}{\to} 0, \tag{3.16}$$

and $\sum_j X_{\epsilon j}$ is asymptotically normal by Theorem 3.6; the variance of $\psi(\epsilon)\sum_j X_{\epsilon j}$ is $\psi^2(\epsilon)\sum_j \mathcal{E}(X_{\epsilon j}^2)$, completing the proof.∎

COROLLARY 3.10: *If $3l - k > 0$ and*

$$s\int \mathcal{E}\Big(|X|^3\Big)(d\mu)^k < \infty, \qquad v = s\int \mathcal{E}(X^2)(d\mu)^{2l} \neq 0,$$

then

$$s\int X(d\mu)^l = N(0, v).$$

§3.3. Distributions Under Symmetric Statistics

Given a spectral measure μ and a spectral μ-net $\mathcal{F}_\epsilon$, the operator A_ϵ with spectrum $\mathcal{F}_\epsilon$ induces a canonical ensemble under the conditions outlined in Chapter 2. For a net of operators $C_\epsilon = g(A_\epsilon)$, (with g a function on $\mathbf{R}^+$) we now consider asymptotics of corresponding value functions c_ϵ; applications will be deferred until later.

DEFINITION 3.11: Let G_β denote the functions of E on $\mathbf{R}^+$ with two continuous derivatives near $x = 0$, locally of bounded variation, such that $|g(E)|(e^{\beta E} + 1)^{-1}$ is non-increasing for E sufficiently large. For $g \in G_\beta$, those non-trivial spectral measures which are 0-free and satisfy

$$0 < \int_0^\infty g^n(E) e^{-\beta E} d\mu(E) < \infty \tag{3.17}$$

for $n = 0, 1, 2, 3$ are denoted by $R_{g,\beta}$. Those $\mu \in R_{g,\beta}$ whose spectral functions have three continuous derivatives near 0 comprise $S_{g,\beta}$.

The monotonicity condition on $|g(E)|(e^{\beta E}+1)^{-1}$ implies the same for $|g(E)|(e^{\beta E}-1)^{-1}$.

The first and second ($n = 0$) conditions on $\mu \in R_{g,\beta}$ guarantee existence of a symmetric canonical ensemble, while the extra one on $S_{g,\beta}$ is required to control asymptotic behavior of c_ϵ. Let $U_{g,\beta}$ denote the spectral measures for which c_ϵ has an asymptotic distribution. Here and in Chapters 4 and 5 we show that $S_{g,\beta} \subset U_{g,\beta}$.

PROPOSITION 3.12: *Let $f(\cdot)$ and $g(\cdot)$ be non-decreasing and non-increasing, respectively, with $g(a)$ possibly infinite, and $f(a) = 0$. Then*

$$0 \leq \int_a^b g(x)df(x) - \int_a^b g(x)d[f(x)] \leq \int_a^{x'} g(x)df(x), \tag{3.18}$$

where

$$x' = \min(\inf\{x : f(x) \geq 1\}, b),$$

b may be ∞, and "a" and "b" may be replaced by "a^-" and "b^-", respectively.

Proof: The proof is the same as that of Lemma 2.5 up to (2.20). We have

$$\sum_{k=1} \int_{S_k} g(x)df(x) \geq \sum_{k=1} \int_{S_k} g(x)d[f(x)] \geq \sum_{k=2} \int_{S_k} g(x)df(x), \tag{3.19}$$

since the fact $f(a) = 0$ implies that

$$\int_{S_k} g(x)df(x) \geq g(x_k)$$

for all k, including $k = 1$. Equation (3.19) immediately gives (3.18) when $b = \infty$. The case of finite b follows similarly, as do the replacements of "a" and "b" by "a^-" and "b^-".∎

PROPOSITION 3.13: *Let $|g(x)|$ be of bounded variation and non-increasing for large x, with f non-decreasing. Then*

$$\int_a^b g(x)d\left[\frac{f(x)}{\epsilon}\right] = \frac{1}{\epsilon}\int_a^b g(x)df(x) + O(1) \qquad (\epsilon \to 0),$$

where b may be infinite, and $a \to a^-$, $b \to b^-$ are allowed.

Proof: Assume $b < \infty$. Since g has bounded variation, it suffices to assume g is monotone. If g is non-increasing, the result follows from Lemma 2.5, the general case following similarly. If $b = \infty$ we may assume that $|g|$ is non-increasing, since any interval in which this fails may be disposed of by the above. In this case Lemma 2.5 again applies, completing the proof.∎

Letting $f(x) = x$ above, we obtain

$$\sum_{k=[\frac{a}{\epsilon}]+1}^{[\frac{b}{\epsilon}]} g(k\epsilon) = \frac{1}{\epsilon}\int_a^b g(x)dx + O(1). \tag{3.20}$$

THEOREM 3.14: *Let $g \in G_\beta$, and $\mu \in S_{g,\beta}$, and one or both of the following hold:*
(i) $F'(0) = F''(0) = 0$; $\mu \in S_{g,\beta}$
(ii) $g(0) = 0$; $\mu \in R_{g,\beta}$,
where F is the spectral function of μ. Let A_ϵ be the ϵ-discrete operator corresponding to μ, and $c_\epsilon = g(A_\epsilon)$. Then the value function c_ϵ of $d\Gamma(C_\epsilon)$ satisfies

$$\sqrt{\epsilon}\Big(c_\epsilon - \frac{m}{\epsilon}\Big) \underset{\epsilon\to 0}{\Rightarrow} N(0,v), \tag{3.21}$$

where

$$m = \int_0^\infty g(E)(e^{\beta E}-1)^{-1}d\mu, \qquad v = \int_0^\infty g(E)e^{\beta E}(e^{\beta E}-1)^{-2}d\mu. \tag{3.22}$$

Proof: (*i*) Let $n_{\epsilon j}$ denote occupation numbers on the symmetric canonical ensemble over A_ϵ; then

$$c_\epsilon = \sum g(E_{\epsilon j})n_{\epsilon j}$$

and $n_{\epsilon j} \sim \gamma(e^{-\beta E_{\epsilon j}})$, where $\gamma(e^{-\beta E})$ denotes the geometric r.v. with parameter $e^{-\beta E}$. Let $\{X(E)\}_{\epsilon > 0}$ denote independent $\gamma(e^{-\beta E})$ r.v.'s, and

$$\hat{X}(E) \equiv X(E) - \mathcal{E}(X(E)). \tag{3.23}$$

Then

$$s\int_a^\infty \mathcal{E}(|g(E)\hat{X}(E)|^3)(d\mu)^{\frac{5}{4}} \leq s\int_0^\infty |g^3(E)|\mathcal{E}(X(E)+\mathcal{E}(X(E)))^3(d\mu)^{\frac{5}{4}}. \tag{3.24}$$

A calculation shows

$$\mathcal{E}\Big\{(X(E)+\mathcal{E}(X(E)))^3\Big\} = \begin{cases} O(e^{-\beta E}) & (E\to\infty) \\ O(E^{-3}) & (E\to 0). \end{cases}$$

Hence the right side of (3.24) is bounded by

$$c_1\, s\int_0^\delta E^{-3}(d\mu)^{\frac{5}{4}} + c_2\, s\int_\delta^\infty e^{-\beta E}|g^3(E)|(d\mu)^{\frac{5}{4}}, \tag{3.25}$$

where $\delta > 0$ is chosen such that (i) δ is not an atom of μ, (ii) for some $K > 0$, $F(E) \leq KE^3$ $(E \leq \delta)$, and (iii) F is continuous for $E \leq \delta$. The second term is 0 by Theorem 3.2. For $E_{\epsilon j} \leq \delta$, $\frac{F(E_{\epsilon j})}{\epsilon} = j$, and thus

$$E_{\epsilon j} \geq \left(\frac{\epsilon j}{K}\right)^{\frac{1}{3}},$$

so that the first term is bounded by

$$\lim_{\epsilon \to 0} c_1 \epsilon^{\frac{5}{4}} \sum_{E_{\epsilon j} \leq \delta} \left(\frac{\epsilon j}{K}\right)^{-1} \leq \lim_{\epsilon \to 0} c_1 K \epsilon^{\frac{1}{4}} \sum_{\left(\frac{\epsilon j}{k}\right)^{\frac{1}{3}} \leq \delta} j^{-1} = \lim_{\epsilon \to 0} \epsilon^{\frac{1}{4}} O(\ln \epsilon) = 0. \tag{3.26}$$

Hence, the left side of (3.24) is 0, while

$$\begin{aligned} 0 < \ s\int_0^\infty \mathcal{E}\Big((g(E)\hat{X}(E))^2\Big) d\mu &= s\int_0^\infty g^2(E)\mathcal{V}(X(E)) d\mu \\ &= s\int_0^\infty g^2(E) e^{\beta E}\left(e^{\beta E} - 1\right)^{-2} d\mu < \infty. \end{aligned}$$

By Corollary 3.10 with $k = \frac{5}{4}$ and $l = \frac{1}{2}$,

$$s\int_0^\infty g(E)\hat{X}(E)(d\mu)^{\frac{1}{2}} = N(0, v), \tag{3.27}$$

with

$$v = \int_0^\infty \frac{g^2(E) e^{\beta E}}{(e^{\beta E} - 1)^2}\ d\mu.$$

Thus c_ϵ is asymptotically normal.

To calculate the asymptotic expectation of c_ϵ, we assume initially that $g(0) > 0$. Then

$$\mathcal{E}(c_\epsilon) = \sum_j \frac{g(E_{\epsilon j})}{(e^{\beta E_{\epsilon j}} - 1)} = \left(\sum_{E_{\epsilon j} \leq \delta} + \sum_{E_{\epsilon j} > \delta}\right) \frac{g(E_{\epsilon j})}{(e^{\beta E_{\epsilon j}} - 1)},$$

with δ chosen so that it is not an atom of μ, and the absolute value of the summand is decreasing in $E_{\epsilon j}$ for $E_{\epsilon j} \leq \delta$. By Proposition (3.13) the second sum on the right is

$$s\int_\delta^\infty \frac{g(E)}{e^{\beta E} - 1} d\left[\frac{F(E)}{\epsilon}\right] = \frac{1}{\epsilon}\int_\delta^\infty \frac{g(E)}{e^{\beta E} - 1}\, d\mu + O(1), \tag{3.28}$$

while

$$\sum_{E_{\epsilon j} \leq \delta} \frac{g(E_{\epsilon j})}{(e^{\beta E_{\epsilon j}} - 1)} = \int_0^\delta \frac{g(E)}{(e^{\beta E} - 1)}\, d\left[\frac{f(E)}{\epsilon}\right] = \frac{1}{\epsilon}\int_0^\delta \frac{g(E)}{(e^{\beta E} - 1)}\, d\mu + h(\epsilon), \tag{3.29}$$

where

$$|h(\epsilon)| \leq \frac{1}{\epsilon} \int_0^{E'(\epsilon)} \frac{|g(E)|}{(e^{\beta E}-1)}\, d\mu,$$

and $E'(\epsilon) \equiv \inf\left\{E' : \frac{f(E')}{\epsilon} \geq 1\right\}$. If F is 0 in a neighborhood of 0 then (3.28) holds for $\delta = 0$. Otherwise, $E'(\epsilon) \underset{\epsilon\to 0}{\longrightarrow} 0$ and

$$|h(\epsilon)| \leq \frac{c}{\epsilon} \int_0^{E_1(\epsilon)} \frac{1}{(e^{\beta E}-1)}\, dF(E) \leq \frac{c}{\epsilon} \int_0^{E_1(\epsilon)} \frac{1}{E} dF(E) = \frac{c}{\epsilon} \int_0^{F(E_1(\epsilon))} \frac{1}{E_1(\epsilon')} d\epsilon',$$

for some $c > 0$ and ϵ sufficiently small. The equality obtains as follows. Let F be thrice continuously differentiable for $E \leq \eta$. $F(\cdot)$ is an inverse of $E_1(\cdot)$, since $F \circ E_1(\epsilon) = \epsilon \quad (0 \leq \epsilon \leq F(\eta))$ (note that F is not generally invertible on $(0, \eta]$, since it may be constant on intervals). Hence equality follows formally by the change of variables $\epsilon' = F(E)$; a rigorous argument involves the definition of the Stieltjes integral as a limit of sums over partitions. Since $F(E) \leq KE^3 \quad (E \in (0, \eta])$,

$$E_1(\epsilon) \geq \left(\frac{\epsilon}{K}\right)^{\frac{1}{3}} \qquad (0 < \epsilon \leq F(\eta)).$$

Hence,

$$\int_0^{F(E_1(\epsilon))} \frac{1}{E_1(\epsilon')} d\epsilon' \leq \frac{3K^{\frac{1}{3}}}{2} F(E_1(\epsilon))^{\frac{2}{3}} = \frac{3K^{\frac{1}{3}}}{2} \epsilon^{\frac{2}{3}}. \tag{3.30}$$

Combining (3.28-30),

$$\mathcal{E}(c_\epsilon) = \frac{1}{\epsilon} \int_0^\infty \frac{g(E)}{e^{\beta E}-1} d\mu + 0(\epsilon^{-\frac{1}{3}}) = \frac{m}{\epsilon} + O\left(\epsilon^{-\frac{1}{3}}\right) \qquad (\epsilon \to 0). \tag{3.31}$$

This calculation is much simpler if $g(0) = 0$ (and hence $g(E) = O(E)$ near zero). By (3.27) and (3.31),

$$\sqrt{\epsilon}\left(c_\epsilon - \frac{m}{\epsilon}\right) = \sqrt{\epsilon}(c_\epsilon - \mathcal{E}(c_\epsilon)) + \sqrt{\epsilon}\left(\mathcal{E}(c_\epsilon) - \frac{m}{\epsilon}\right)$$

$$\underset{\epsilon\to 0}{\Rightarrow} s \int_0^\infty g(E)\hat{X}(E)(d\mu)^{\frac{1}{2}} + 0 = N(0, v),$$

completing the proof of (i).

(ii) In this case $g(E) = O(E) \quad (E \to 0)$, and

$$\mathcal{E}(|g(E)\hat{X}(E)|^3) = \begin{cases} O(e^{-\beta E}); & (E \to \infty) \\ O(1); & (E \to 0) \end{cases} \tag{3.32}$$

with the same bounds for $\mathcal{E}(g^2(E)\hat{X}^2(E))$. Hence asymptotic normality of c_ϵ follows by Corollary 3.10, with $k = 1$, $l = \frac{1}{2}$. By Proposition 3.13,

$$\mathcal{E}(c_\epsilon) = \int_0^\infty \frac{g(E)}{e^{\beta E} - 1} d\left[\frac{F(E)}{\epsilon}\right] = \frac{1}{\epsilon}\int_0^\infty \frac{g(E)}{e^{\beta E} - 1} d\mu + O(1);$$

$$V(c_\epsilon) = \frac{1}{\epsilon}\int_0^\infty \frac{g(E)e^{\beta E}}{(e^{\beta E} - 1)^2} d\mu + O(1),$$

completing the proof.∎

§ 3.4. Antisymmetric Distributions

DEFINITION 3.15: The class H_β is defined in the same way as G_β (Def. 3.11), without the condition on differentiability. For $g \in H_\beta$, $T_{g,\beta}$ is $R_{g,\beta}$ without the 0-freedom condition.

The relaxed conditions above still allow proof of central limit theorems for observables under antisymmetric statistics, since antisymmetric occupation numbers form a non-singular family. In this case $n_{\epsilon j}$ is Bernoulli (see §2.2), with moments easily bounded by

$$\mathcal{E}\left(|\hat{n}_{\epsilon j}|^k\right) \le \mathcal{E}\left(\left(n_{\epsilon j} + \left(e^{\beta E_{\epsilon j}} + 1\right)^{-1}\right)^k\right) = O\left(e^{-\beta E_{\epsilon j}}\right) \qquad (k = 0, 1, 2, \ldots) \tag{3.33}$$

where

$$\hat{n}_{\epsilon j} \equiv n_{\epsilon j} - \mathcal{E}(n_{\epsilon j}) = n_{\epsilon j} - \left(e^{\beta E_{\epsilon j}} + 1\right)^{-1}.$$

By Corollary 3.10, $c_\epsilon = \sum g(E_{\epsilon j}) n_{\epsilon j}$ is asymptotically normal. By Proposition 3.13,

$$\begin{aligned} \mathcal{E}(c_\epsilon) &= \int_0^\infty \left(\frac{g(E)}{e^{\beta E} + 1}\right) d\left[\frac{F(E)}{\epsilon}\right] = \frac{1}{\epsilon}\int_0^\infty \left(\frac{g(E)}{e^{\beta E} + 1}\right) d\mu + O(1) \\ V(c_\epsilon) &= \frac{1}{\epsilon}\int_0^\infty \frac{g(E)e^{\beta E}}{(e^{\beta E} + 1)^2} d\mu + O(1) \qquad (\epsilon \to 0). \end{aligned} \tag{3.34}$$

Applying Proposition 3.13 to (3.34), we have

THEOREM 3.16: *If $g \in H_\beta$, and $\mu \in T_{g,\beta}$, then*

$$\sqrt{\epsilon}\left(c_\epsilon - \frac{m}{\epsilon}\right) \Rightarrow N(0, v),$$

where

$$m = \int_0^\infty \frac{g(E)}{(e^{\beta E} + 1)} d\mu, \qquad v = \int_0^\infty \frac{g(E)e^{\beta E}}{(e^{\beta E} + 1)^2} d\mu.$$

CHAPTER 4

SINGULAR CENTRAL LIMIT THEOREMS, I

The integration theory of Chapter 3 deals with non-singular integrals in that, e.g., $_S \int \mathcal{E}(X^2) d\mu$ is finite. This section, from a probabilistic standpoint, is an illustration of the singular theory for the family of geometric r.v.'s arising under symmetric statistics. The non-normality of these integrals, demonstrated in Chapter 5, is closely connected to the "infrared catastrophe" of quantum electrodynamics.

The most convenient formulation (with a view toward applications) is in the language of limits rather than integrals. By Theorem 3.14, we may assume that $g(0) > 0$.

§4.1. The Case $F(E) = aE^2$, $g(E) \equiv 1$

With the assumptions of §3.3 still in force, this falls into the category of "weakest" violations of the hypotheses of Theorem 3.14. We have

$$E_{\epsilon j} = \sqrt{\frac{\epsilon j}{a}}; \qquad V(c_\epsilon) = a(\beta\phi(\epsilon))^{-2} + O\left(\frac{1}{\epsilon}\right), \tag{4.1}$$

where

$$\phi(\epsilon) = \sqrt{\frac{\epsilon}{|\ln \epsilon|}}$$

(a definition which will hold henceforth); hence Theorem 3.14 cannot hold here. The asymptotic distribution of c_ϵ, though normal, has a new nature in that, after normalization to standard variance and mean, c_ϵ is asymptotically dominated by occupation numbers $n_{\epsilon j}$ whose spectral values $E_{\epsilon j}$ are arbitrarily small. Normality is in fact not universal in this situation, as will be seen later.

Note that in all "singular" sums c_ϵ of r.v.'s to be considered, leading asymptotics will depend only on spectral behavior near the origin; parameters related to global spectral density will not contribute. We now consider a prototype.

LEMMA 4.1: *If $F(E) = aE^2$ ($E \geq 0$), and $g(E) = 1$, then*

$$R_\epsilon \equiv \phi(\epsilon)\left(c_\epsilon - \frac{\pi^2}{3\epsilon\beta^2}\right) \Rightarrow N\left(0, \frac{a}{\beta^2}\right). \tag{4.2}$$

Proof: Assume without loss that $a = 1$. The characteristic function of R_ϵ is

$$\Phi_{R_\epsilon}(t) = \mathcal{E}\left(e^{iR_\epsilon t}\right) = e^{\left(-it\frac{\phi(\epsilon)\pi^2}{3\epsilon\beta^2}\right)}\mathcal{E}\left(e^{it\phi(\epsilon)c_\epsilon}\right)$$

$$= e^{\left(-it\frac{\phi(\epsilon)\pi^2}{3\epsilon\beta^2}\right)}\prod_{k=1}^{\infty}\left\{\frac{1-e^{-\beta\sqrt{\epsilon k}}}{1-e^{(-\beta\sqrt{\epsilon k}+i\phi(\epsilon)t)}}\right\},$$

so that

$$\ln\Phi_{R_\epsilon}(t) = \frac{-it\phi(\epsilon)\pi^2}{3\epsilon\beta^2} + \sum_{k=1}^{\infty}\left\{\ln\left(1-e^{-\beta\sqrt{\epsilon k}}\right) - \ln\left(1-e^{(-\beta\sqrt{\epsilon k}+i\phi(\epsilon)t)}\right)\right\}. \tag{4.3}$$

The branch of the logarithm is determined by analytic continuation from the real axis for k large. We rewrite (4.3) as

$$\ln\Phi_{R_\epsilon}(t) = \frac{-it\phi(\epsilon)\pi^2}{3\epsilon\beta^2}$$

$$+\sum_{k=1}^{[\frac{1}{\epsilon}]}\left(\ln\left(1-e^{-\beta\sqrt{\epsilon k}}\right)-\ln(\beta\sqrt{\epsilon k})\right)$$

$$-\sum_{k=1}^{[\frac{1}{\epsilon}]}\left(\ln\left(1-e^{(-\beta\sqrt{\epsilon k}+i\phi(\epsilon)t)}\right)-\ln(\beta\sqrt{\epsilon k}-i\phi(\epsilon)t)\right)$$

$$+\sum_{k=1}^{[\frac{1}{\epsilon}]}\left\{\ln(\beta\sqrt{\epsilon k})-\ln(\beta\sqrt{\epsilon k}-i\phi(\epsilon)t)\right\}$$

$$+\sum_{k=[\frac{1}{\epsilon}+1]}^{\infty}\left\{\ln\left(1-e^{-\beta\sqrt{\epsilon k}}\right)-\ln\left(1-e^{(-\beta\sqrt{\epsilon k}+i\phi(\epsilon)t)}\right)\right\}$$

$$=\sum_{k=1}^{[\frac{1}{\epsilon}]}\left\{h(\beta\sqrt{\epsilon k})-h(\beta\sqrt{\epsilon k}-i\phi(\epsilon)t)\right\}$$

$$+\sum_{k=1}^{[\frac{1}{\epsilon}]}\left\{\ln(\beta\sqrt{\epsilon k})-\ln(\beta\sqrt{\epsilon k}-i\phi(\epsilon)t)\right\} \tag{4.4}$$

$$+\sum_{k=[\frac{1}{\epsilon}+1]}^{\infty}\left\{l(\beta\sqrt{\epsilon k})-l(\beta\sqrt{\epsilon}k-i\phi(\epsilon)t)\right\},$$

where

$$h(x)=\ln\left(\frac{1-e^{-x}}{x}\right);\qquad l(x)=\ln(1-e^{-x}).$$

In the first sum ϵk ranges from 0 to 1, so that

$$\begin{aligned}
&\sum_{k=1}^{[\frac{1}{\epsilon}]}\left\{h(\beta\sqrt{\epsilon k}) - h(\beta\sqrt{\epsilon k} - i\phi(\epsilon)t)\right\} \\
&= \left\{\sum_{k=1}^{[\frac{1}{\epsilon}]} h'(\beta\sqrt{\epsilon k})(i\phi(\epsilon)t)\right\}(1 + O(\phi(\epsilon))) \qquad (\epsilon \to 0).
\end{aligned} \tag{4.5}$$

Similarly

$$\begin{aligned}
&\sum_{k=[\frac{1}{\epsilon}+1]}^{\infty}\left\{l(\beta\sqrt{\epsilon k}) - \ln(\beta\sqrt{\epsilon k} - i\phi(\epsilon)t)\right\} \\
&= \left\{\sum_{k=[\frac{1}{\epsilon}]+1}^{\infty} l'(\beta\sqrt{\epsilon k})i\phi(\epsilon)t\right\}(1 + O(\phi(\epsilon))) \qquad (\epsilon \downarrow 0)
\end{aligned} \tag{4.6}$$

Note that $h'(\beta\sqrt{x}) = \frac{\beta\sqrt{x}+1-e^{\beta\sqrt{x}}}{\beta\sqrt{x}(e^{\beta\sqrt{x}}-1)}$ is bounded and monotone decreasing for $x \in [0,1]$, as is $l'(\beta\sqrt{x}) = \frac{1}{e^{\beta\sqrt{x}}-1}$, for $x \in [1,\infty)$. Thus (3.20) applies to the sums in (4.5) and (4.6), and (4.4) becomes

$$\begin{aligned}
\ln \Phi_{R_\epsilon}(t) &= \frac{-it\phi(\epsilon)\pi^2}{3\epsilon\beta^2} + \left\{\frac{1}{\epsilon}\int_0^1 h'(\beta\sqrt{x})dx + O(1)\right\}(1 + O(\phi(\epsilon)))i\phi(\epsilon)t \\
&\quad + \left\{\frac{1}{\epsilon}\int_1^{\infty} l'(\beta\sqrt{x})dx + O(1)\right\}(1 + O(\phi(\epsilon)))i\phi(\epsilon)t \\
&\quad + \sum_{k=1}^{[\frac{1}{\epsilon}]}\left\{\ln(\beta\sqrt{\epsilon k}) - \ln(\beta\sqrt{\epsilon k} - i\phi(\epsilon)t)\right\} \\
&= \frac{-it\phi(\epsilon)\pi^2}{3\epsilon\beta^2} + \frac{i\phi(\epsilon)t}{\epsilon}\left\{\int_0^{\infty}\frac{1}{e^{\beta\sqrt{x}}-1}dx - \int_0^1 \frac{1}{\beta\sqrt{x}}dx\right\}(1 + O(\phi(\epsilon))) \\
&\quad - \sum_{k=1}^{[\frac{1}{\epsilon}]}\ln\left(1 - \frac{i\phi(\epsilon)t}{\beta\sqrt{\epsilon k}}\right) \qquad (\epsilon \to 0).
\end{aligned} \tag{4.7}$$

For small ϵ the arguments of all logarithms are near $\mathbf{R}^+ \subset \mathbf{C}$, so that branches are defined by analytic continuation.

For $x \in \mathbf{C}$ and $|x| \le \frac{1}{2}$, $\ln(1-x) = -x - \frac{x^2}{2}(1 + xS(x))$, where $S(\cdot)$ is bounded. Noting that

$$\frac{\phi(\epsilon)t}{\beta\sqrt{\epsilon k}} \le \frac{\phi(\epsilon)t}{\beta\sqrt{\epsilon}} \underset{\epsilon\to 0}{\to} 0 \tag{4.8}$$

we have

$$\sum_{k=1}^{[\frac{1}{\epsilon}]} \ln\left(1-\frac{i\phi(\epsilon)t}{\beta\sqrt{\epsilon k}}\right) = -\sum_{k=1}^{[\frac{1}{\epsilon}]}\left(\frac{i\phi(\epsilon)t}{\beta\sqrt{\epsilon k}}\right) + \left(\sum_{k=1}^{[\frac{1}{\epsilon}]}\frac{\phi(\epsilon)^2t^2}{2\beta^2\epsilon k}\right)(1+o(1))$$

$$= \frac{-i\phi(\epsilon)t}{\beta\sqrt{\epsilon}}\sum_{k=1}^{[\frac{1}{\epsilon}]}\left(\frac{1}{\sqrt{k}}\right) + \frac{\phi(\epsilon)^2t^2}{2\epsilon\beta^2}\left(\sum_{k=1}^{[\frac{1}{\epsilon}]}\frac{1}{k}\right)(1+o(1)) \tag{4.9}$$

$$= \frac{-i\phi(\epsilon)t}{\beta\sqrt{\epsilon}}\left(2\sqrt{\left[\frac{1}{\epsilon}\right]}+O(1)\right) + \frac{\phi(\epsilon)^2t^2}{2\epsilon\beta^2}\left(\ln\left[\frac{1}{\epsilon}\right]+O(1)\right)(1+o(1)) \qquad (\epsilon\to 0).$$

Finally, replacing $[\frac{1}{\epsilon}]$ by $\frac{1}{\epsilon}$ we have

$$\sum_{k=1}^{[\frac{1}{\epsilon}]} \ln\left(1-\frac{\phi(\epsilon)t}{\beta\sqrt{\epsilon k}}\right) = \frac{-2i\phi(\epsilon)t}{\beta\epsilon} + \frac{t^2}{2\beta^2} + o(1) \qquad (\epsilon\to 0),$$

Thus

$$\ln\Phi_{R_\epsilon}(t) = -\frac{it\phi(\epsilon)\pi^2}{3\epsilon\beta^2} + \frac{i\phi(\epsilon)t}{\epsilon}\left\{\frac{\pi^2}{3\beta^2}-\frac{2}{\beta}\right\} + \frac{2i\phi(\epsilon)t}{\beta\epsilon} - \frac{t^2}{2\beta^2} + o(1)$$

$$= -\frac{t^2}{2\beta^2} + o(1) \quad (\epsilon\to 0). \tag{4.10}$$

Since $e^{-\frac{t^2}{\beta^2}}$ is the characteristic function of $N\left(0,\frac{1}{\beta^2}\right)$, the Lévy-Cramér convergence theorem completes the proof.∎

§4.2 The Case $(gF')'(0) > 0,\ gF'(0) = 0$

In this case the "leading behavior" of c_ϵ arises from its small $E_{\epsilon j}$ terms and can be analyzed with use of the previous lemma. Recall that the convolution of distribution functions is defined by

$$(F_1 * F_2)(x) = \int F_1(x-y)dF_2(y), \tag{4.11}$$

with the obvious extension to probability laws.

LEMMA 4.2: *Let $\{X_i\}_{i=1}^{\infty}$ be a collection of independent r.v.'s and $\{X'_i\}_{i=1}^{\infty}$ be another. Let $S = \sum X_j$ and $S' = \sum X'_j$ converge a.e. If $F_{X_i} \le F_{X'_i}(x)$ for all x, then $F_S(x) \le F_{S'}(x)$.*

Proof: This follows by direct calculation for finite sums, and in the general case from the fact that almost sure implies distributional convergence.∎

The sets G_β and $S_{g,\beta}$ are defined in Def. 3.11.

THEOREM 4.3: *Let $g \in G_\beta$ and $\mu \in S_{g,\beta}$, with $g(0) > 0$ and F the spectral function of μ. Assume that $F'(0) = 0$ and $F''(0) = 2a > 0$. Then*

$$\phi(\epsilon)\Big(c_\epsilon - \frac{m}{\epsilon}\Big) \Rightarrow N\Big(0, \frac{ag(0)}{\beta^2}\Big),$$

with m given by (3.22).

Proof: We prove this in three parts, first assuming $g \equiv 1$, $F'''(0^+) < 0$, and then sequentially eliminating the second and first conditions.

I. Assume

$$F'''(0^+) < 0, \qquad g(E) \equiv 1.$$

Then

$$c_\epsilon = n_\epsilon = \sum_{j=1}^{\infty} n_{\epsilon j},$$

and since $\mu \in S_{g,\beta}$,

$$F'''(E) < 0 \qquad (0 \leq E \leq \lambda) \tag{4.12}$$

for some $\lambda > 0$. Let

$$F^\lambda(E) = \begin{cases} F(E); & E \leq \lambda \\ F(\lambda); & E \geq \lambda \end{cases}, \tag{4.13}$$

μ^λ be the spectral measure defined by $F^\lambda(\cdot)$, and $n_\epsilon^\lambda = \sum_j n_{\epsilon j}^\lambda$ the corresponding total occupation number. Define

$$H_\epsilon(E) = \left[\frac{F^\lambda(E)}{\epsilon}\right] + \left[\frac{aE^2 - F^\lambda(E)}{\epsilon}\right], \tag{4.14}$$

and let $\mathcal{F}_\epsilon^* = \{E_{\epsilon j}^*\}_{1 \leq j < \infty}$ be the ordered discontinuities of $H_\epsilon(\cdot)$, the size of a discontinuity being its multiplicity in $\mathcal{F}_\epsilon^*$. Let $\mathcal{F}_\epsilon^a = \{E_{\epsilon j}^a\}_{1 \leq j < \infty}$ be the discontinuities of $[\frac{aE^2}{\epsilon}]$, so that

$$E_{\epsilon j}^a = \sqrt{\frac{\epsilon j}{a}}. \tag{4.15}$$

Let A_ϵ^* and A_ϵ^a be self adjoint operators in Hilbert space with spectra $\mathcal{F}_\epsilon^*$ and $\mathcal{F}_\epsilon^a$, respectively, again with multiplicity. Note that

$$0 \leq \left[\frac{aE^2}{\epsilon}\right] - H_\epsilon(E) \leq 1 \qquad (E \in \mathbf{R}), \tag{4.16}$$

since for $a, b \in \mathbf{R}$,

$$0 \leq [a + b] - \{[a] + [b]\} \leq 1.$$

Furthermore, if $E^*_{\epsilon(j-1)} > E^a_{\epsilon j}$, then, for $E^a_{\epsilon j} < E < E^*_{\epsilon(j-1)}$

$$\left[\frac{aE^2}{\epsilon}\right] \geq j, \qquad H_\epsilon(E) < j-1,$$

contradicting (4.16). Thus,

$$E^*_{\epsilon(j-1)} \leq E^a_{\epsilon j} \qquad (2 \leq j < \infty). \tag{4.17}$$

Also, by (4.16),

$$E^a_{\epsilon j} \leq E^*_{\epsilon j} \qquad (1 \leq j < \infty). \tag{4.18}$$

Let $\{n^*_{\epsilon j}\}_{1 \leq j < \infty}$ and $\{n^a_{\epsilon j}\}_{1 \leq j < \infty}$ be the symmetric occupation number r.v.'s corresponding to A^*_ϵ and A^a_ϵ, and

$$n^*_\epsilon = \sum_{j=1}^{\infty} n^*_{\epsilon j}, \qquad n^a_\epsilon = \sum_{j=1}^{\infty} n^a_{\epsilon j} \tag{4.19}$$

total number r.v.'s. By Lemma 4.1 and the definition of A^a_ϵ,

$$Q^a_\epsilon \equiv \frac{\beta\phi(\epsilon)}{\sqrt{a}}\left(n^a_\epsilon - \frac{a\pi^2}{3\epsilon\beta^2}\right) \underset{\epsilon\to 0}{\Rightarrow} N(0,1). \tag{4.20}$$

Note also that by (4.15),

$$\begin{aligned}
\mathcal{E}\left\{\left(\frac{\beta\phi(\epsilon)}{\sqrt{a}} n^a_{\epsilon 1}\right)^2\right\} &= \frac{\beta^2\phi(\epsilon)^2}{a}\left\{\mathcal{V}(n^a_{\epsilon 1}) + \mathcal{E}\left((n_{\epsilon 1})^2\right)\right\} \\
&= \frac{\beta^2\phi(\epsilon)^2}{a}\left(\frac{\exp\left(\beta\sqrt{\frac{\epsilon}{a}}\right)}{\left(\exp\left(\beta\sqrt{\frac{\epsilon}{a}}\right)-1\right)^2} + \frac{1}{\left(\exp\left(\beta\sqrt{\frac{\epsilon}{a}}\right)-1\right)^2}\right) \\
&= O\left(\frac{1}{|\ln\epsilon|}\right) \underset{\epsilon\to 0}{\to} 0,
\end{aligned}$$

so that

$$W^a_\epsilon \equiv \frac{\beta\phi(\epsilon)}{\sqrt{a}} n^a_{\epsilon 1} \underset{\epsilon\to 0}{\Rightarrow} \delta_0, \tag{4.21}$$

with δ_0 the point mass at 0. If we define

$$n^{a,\bar{1}}_\epsilon = n^a_\epsilon - n_{\epsilon 1},$$

then (4.21) and (4.20) imply

$$Q^{a,1}_\epsilon \equiv Q^a_\epsilon - W^a_\epsilon = \frac{\beta\phi(\epsilon)}{\sqrt{G}}\left(n^{a,1}_\epsilon - \frac{a\pi^2}{3\epsilon\beta^2}\right) \underset{\epsilon\to 0}{\Rightarrow} N(0,1). \tag{4.22}$$

The d.f. of $n^a_{\epsilon j}$ is

$$F_{n^a_{\epsilon j}}(x) = \sum_{k=0}^{[x]} \mathcal{P}(n_{\epsilon j} = k) = \sum_{k=0}^{[x]} \left(1 - e^{-\beta E^a_{\epsilon j}}\right) e^{-\beta k E^a_{\epsilon j}} = 1 - e^{-\beta [x+1] E^a_{\epsilon j}} \quad (x \in \mathbf{R}),$$

which is monotonic in $E^a_{\epsilon j}$. By (4.17) and (4.18),

$$\left.\begin{aligned} F_{n^*_{\epsilon(j-1)}}(x) &\le F_{n^a_{\epsilon j}}(x) \qquad (j \ge 2) \\ F_{n^a_{\epsilon j}}(x) &\le F_{n^*_{\epsilon j}}(x) \qquad (j \ge 1) \end{aligned}\right\}. \tag{4.23}$$

We decompose

$$\mathcal{F}^*_\epsilon = \mathcal{F}^{*I}_\epsilon \cup \mathcal{F}^{*II}_\epsilon,$$

where the set $\mathcal{F}^{*I}_\epsilon$ corresponds to the discontinuities in $\left[\frac{F^\lambda(E)}{\epsilon}\right]$ and $\mathcal{F}^{*II}_\epsilon$ to those in $\left[\frac{aE^2 - F^\lambda(E)}{\epsilon}\right]$; the intersection of $\mathcal{F}^{*I}_\epsilon$ and $\mathcal{F}^{*II}_\epsilon$ is listed twice in $\mathcal{F}^*_\epsilon$. Let

$$S^I_\epsilon = \{j : E^*_{\epsilon j} \in \mathcal{F}^{*I}_\epsilon\}, \qquad S^{II}_\epsilon = \{j : E^*_{\epsilon j} \in \mathcal{F}^{*II}_\epsilon\}, \tag{4.24}$$

and define

$$n^{*I}_\epsilon = \sum_{j \in S^I_\epsilon} n^*_{\epsilon j}, \qquad n^{*II}_\epsilon = \sum_{j \in S^{II}_\epsilon} n^*_{\epsilon j}. \tag{4.25}$$

Note that n^{*I}_ϵ (n^{*II}_ϵ) is the total occupation number corresponding to the d.f.

$$F^{*I}_\epsilon(E) = \frac{F^\lambda(E)}{\epsilon} \qquad \left(F^{*II}_\epsilon(E) = \frac{1}{\epsilon}(aE^2 - F^\lambda(E))\right).$$

By their definitions, n^λ_ϵ and n^{*I}_ϵ, are identically distributed.

By (4.12-13), $F^{*II}(\cdot)$ is monotone non-decreasing. Since $\mu \in S_{1,\beta}$,

$$F^{*II} \in S_{1\beta},$$

and

$$(F^{*II})'(0^+) = (F^{*II})''(0^+) = 0. \tag{4.26}$$

By Theorem 3.14 and the above,

$$Q^{*II}_\epsilon \equiv \sqrt{\frac{\epsilon}{v^{*II}}}\left(n^{*II}_\epsilon - \frac{m^{*II}}{\epsilon}\right) \Rightarrow N(0,1),$$

where

$$m^{*II} = \int_0^\infty \frac{1}{e^{\beta E} - 1} d(aE^2 - F^\lambda(E)),$$

$$v^{*II} = \int_0^\infty \frac{e^{\beta E}}{(e^{\beta E} - 1)^2} d(aE^2 - F^\lambda(E)).$$

The r.v.

$$n_\epsilon^{a,1} = \sum_{j=2}^{\infty} n_{\epsilon j}^a. \tag{4.27}$$

has l^{th} term $n^a_{\epsilon(l-1)}$ while the lth term of n^* in (4.19) is $n^*_{\epsilon l}$. Thus by (4.23) the d.f. of the lth term of $n_\epsilon^{a,1}$ is bounded from above by that of the corresponding term in $n^*_{\epsilon l}$. Hence by Lemma 4.2, letting $F_n(x)$ denote the d.f. of n,

$$F_{n_\epsilon^*}(x) \leq F_{n_\epsilon^{a,1}}(x),$$

and similarly, by (4.24),

$$F_{n_\epsilon^a}(x) \leq F_{n_\epsilon^*}(x).$$

Since these inequalities are preserved under renormalization of expectations and variances, we have by (4.20)

$$F_{Q_\epsilon^a}(x) \leq F_{Q_\epsilon^*}(x); \qquad F_{Q_\epsilon^*}(x) \leq F_{Q_\epsilon^{a,1}}(x), \tag{4.28}$$

where

$$Q_\epsilon^* = \frac{\beta\phi(E)}{\sqrt{a}}\left(n_\epsilon^* - \frac{a\pi^2}{3\epsilon\beta^2}\right).$$

Thus by (4.28), and (4.22),

$$F_{Q_\epsilon^a}(x) \leq F_{Q_\epsilon^*}(x) \leq F_{Q_\epsilon^{a,1}}(x) \underset{\epsilon\to 0}{\to} F_{N(0,1)}(x), \tag{4.29}$$

and by (4.29),

$$Q_\epsilon^* \Rightarrow N(0,1). \tag{4.30}$$

But by the definition of Q_ϵ^*,

$$\begin{aligned} Q_\epsilon^* &= \frac{\beta\phi(\epsilon)}{\sqrt{a}}\left(n_\epsilon^{*I} - \frac{a\pi^2 - 3\beta^2 m^{*II}}{3\epsilon\beta^2}\right) + \frac{\beta\phi(\epsilon)}{\sqrt{a}}\left(n^{*II} - \frac{m^{*II}}{\epsilon}\right) \\ &= \frac{\beta\phi(\epsilon)}{\sqrt{a}}\left(n_\epsilon^{*I} - \frac{a\pi^2 - 3\beta^2 m^{*II}}{3\epsilon\beta^2}\right) + \frac{\beta\sqrt{v^{*II}}\phi(\epsilon)}{\sqrt{a\epsilon}} Q_\epsilon^{*II}. \end{aligned} \tag{4.31}$$

The coefficient of Q_ϵ^{*II} approaches 0 and so the second term converges in law to δ_0. Thus by (4.31), (4.30), the independence of n_ϵ^{*I} and n_ϵ^{*II}, and the identity of n_ϵ^{*I} and n_ϵ^λ,

$$\frac{\beta\phi(\epsilon)}{\sqrt{a}}\left(n_\epsilon^\lambda - \frac{a\pi^2 - 3\beta^2 m^{*II}}{3\epsilon\beta^2}\right) \underset{\epsilon\to 0}{\Rightarrow} N(0,1).$$

Now let

$$n_\epsilon^> = n_\epsilon - n_\epsilon^\lambda. \tag{4.32}$$

Using by now standard arguments, we conclude

$$\sqrt{\frac{\epsilon}{v^{>}(\lambda)}}\left(n_\epsilon^{>}(\lambda) - \frac{m^{>}(\lambda)}{\epsilon}\right) \Rightarrow N(0,1), \tag{4.33}$$

where

$$m^{>}(\lambda) = \int_\lambda^\infty \frac{1}{e^{\beta E} - 1} dF(E); \qquad v^{>}(\lambda) = \int_\lambda^\infty \frac{e^{\beta E}}{(e^{\beta E} - 1)^2} dF(E).$$

Equation (4.32) implies

$$\begin{aligned}\frac{\beta\phi(\epsilon)}{\sqrt{a}}\left(n_\epsilon - \frac{m}{\epsilon}\right) &= \frac{\beta\phi(\epsilon)}{\sqrt{a}}\left(n_\epsilon^{\lambda}(\lambda) - \frac{a\pi^2 - 3\beta^2 m^{*II}}{3\epsilon\beta^2}\right) \\ &+ \frac{\beta\phi(\epsilon)}{\sqrt{a}}\left(n_\epsilon^{>}(\lambda) - \left(\frac{m^{*II}}{\epsilon} + \frac{m}{\epsilon} - \frac{a\pi^2}{3\epsilon\beta^2}\right)\right).\end{aligned} \tag{4.34}$$

By the definitions,

$$m^{*II} + m - \frac{a\pi^2}{3\beta^2} = \int_\lambda^\infty \frac{1}{e^{\beta E} - 1} dF(E) = m^{>}(\lambda), \tag{4.35}$$

since

$$\int_0^\infty \frac{1}{e^{\beta E} - 1} d(aE^2) = \frac{a\pi^2}{3\beta^2}.$$

Because $\frac{\phi(\epsilon)}{\sqrt{\epsilon}} \to 0$, the last term in (4.34) converges to δ_0 by (4.33) and (4.35). On the other hand, the first term on the right of (4.34) converges to the standard normal, so that

$$\frac{\beta\phi(\epsilon)}{\sqrt{a}}\left(n_\epsilon - \frac{m}{\epsilon}\right) \underset{\epsilon\to 0}{\Rightarrow} N(0,1) \tag{4.36}$$

completing part I.

II. We now eliminate the restriction on $F'''(0)$ and assume only that $g(E) \equiv 1$. In this case there clearly exists a spectral function $F^0(\cdot)$ corresponding to a distribution $\mu^0 \in S_{1\beta}$ such that (1) $F(E) - F^0(E)$ is monotone non-decreasing for $E \in \mathbf{R}$, (2) $(F^0)'(0^+) = 0$, (3) $(F^0)''(0^+) = a$, and (4) $(F^0)'''(0^+) < 0$. The construction of such a spectral function involves finding one with sufficiently negative third derivative which is constant for all values of E larger than some sufficiently small number $E_1 > 0$. The function F^0 satisfies the hypotheses of Part I.

Analogously to (4.14), define

$$F_\epsilon^* = \left[\frac{F^0(E)}{\epsilon}\right] + \left[\frac{F(E) - F^0(E)}{\epsilon}\right]. \tag{4.37}$$

Let $\mathcal{F}_\epsilon^* = \{E_{\epsilon j}^*\}_{1 \leq j < \infty}$ be the ordered set of discontinuities of F_ϵ^* and $\mathcal{F}_\epsilon = \{E_{\epsilon j}\}_{1 \leq j < \infty}$ so those of $\left[\frac{F(E)}{\epsilon}\right]$. As before

$$0 \leq \left[\frac{F(E)}{\epsilon}\right] - F_\epsilon^*(E) \leq 1 \qquad (E \in \mathbf{R}),$$

so that

$$E_{\epsilon j} \leq E_{\epsilon j}^*; \qquad E_{\epsilon (j-1)}^* \leq E_{\epsilon j}. \tag{4.38}$$

Let

$$n_\epsilon = \sum_{j=1}^{\infty} n_{\epsilon j}, \qquad n_\epsilon^* = \sum_{j=1}^{\infty} n_{\epsilon j}^* \tag{4.39}$$

be the total symmetric number r.v.'s obtained from $F(E)$ and $F^*(E)$, respectively, and $n_\epsilon^1 = n_\epsilon - n_{\epsilon 1}$. As before we conclude from (4.38) that

$$F_{R_\epsilon}(x) \leq F_{R_\epsilon^*}(x); \qquad F_{R_\epsilon^*}(x) \leq F_{R_\epsilon^1}(x), \tag{4.40}$$

where

$$R_\epsilon = \frac{\phi(\epsilon)\beta}{\sqrt{a}}\left(n_\epsilon - \frac{m}{\epsilon}\right),$$

with R_ϵ^* and R_ϵ^1 corresponding similarly to n_ϵ^* and n_ϵ^1, respectively.

We make the decomposition $\mathcal{F}_\epsilon^* = \mathcal{F}_\epsilon^{*I} \cap \mathcal{F}_\epsilon^{*II}$, where $\mathcal{F}_\epsilon^{*I}$ corresponds to discontinuities in $\left[\frac{F^0(E)}{\epsilon}\right]$ and $\mathcal{F}_\epsilon^{*II}$ to those in $\left[\frac{F(E)-F^0(E)}{\epsilon}\right]$. Let n_ϵ^{*I} and n_ϵ^{*II} be defined analogously to (4.25). By Theorem 3.14,

$$\sqrt{\frac{\epsilon}{v^{*II}}}\left(n_\epsilon^{*II} - \frac{m^{*II}}{\epsilon}\right) \Rightarrow N(0,1), \tag{4.41}$$

with

$$\begin{aligned} v^{*II} &= \int_0^\infty \frac{e^{\beta E}}{(e^{\beta E}-1)^2} d(F(E) - F^0(E)); \\ m^{*II} &= \int_0^\infty \frac{1}{e^{\beta E}-1} d(F(E) - F^0(E)). \end{aligned} \tag{4.42}$$

By the conclusion of Part I,

$$\frac{\beta\phi(\epsilon)}{\sqrt{a}}\left(n_\epsilon^{*I} - \frac{m^{*I}}{\epsilon}\right) \Rightarrow N(0,1),$$

where

$$m^{*I} = \int_0^\infty \frac{1}{e^{\beta E}-1} dF^0(E).$$

Hence,

$$R_\epsilon^* \Rightarrow N(0,1). \tag{4.43}$$

As in (4.21),

$$W_\epsilon \equiv \frac{\phi(\epsilon)\beta}{\sqrt{a}} n_{\epsilon 1} \Rightarrow \delta_0. \tag{4.44}$$

Since $F_{N(0,1)}(\cdot)$ is continuous, a net of probability distributions which converges to $F_{N(0,1)}(\cdot)$ pointwise does so uniformly in x. Thus by (4.44),

$$s_\epsilon \equiv \sup_{x \in \mathbf{R}} |F_{N(0,1)}(x) - F_{R_\epsilon^*}(x)| \underset{\epsilon \to 0}{\to} 0,$$

and

$$\begin{aligned} |F_{R_\epsilon^*} * F_{W_\epsilon}(x) - F_{N(0,1)} * F_{W_\epsilon}(x)| &= \int_{-\infty}^{\infty} (F_{R_\epsilon^*}(x-y) - F_{N(0,1)}(x-y)) dF_{W_\epsilon}(y) \\ &\leq s_\epsilon \int_{-\infty}^{\infty} dF_{W_\epsilon}(y) = s_\epsilon \underset{\epsilon \to 0}{\to} 0. \end{aligned} \tag{4.45}$$

Also by (4.44),

$$F_{N(0,1)} * F_{W_\epsilon}(x) \underset{\epsilon \to 0}{\to} F_{N(0,1)}(x),$$

so that

$$F_{R_\epsilon^*} * F_{W_\epsilon}(x) \underset{\epsilon \to 0}{\to} F_{N(0,1)}(x). \tag{4.46}$$

Thus Lemma 4.2, (4.44), (4.40), and (4.43) imply

$$F_{R_\epsilon^*} * F_{W_\epsilon}(x) \leq F_{R_\epsilon^1} * F_{W_\epsilon}(x) = F_{R_\epsilon}(x) \leq F_{R_\epsilon^*}(x) \underset{\epsilon \to 0}{\to} F_{N(0,1)}(x) \qquad (x \in \mathbf{R}),$$

so that, with (4.46), we have

$$R_\epsilon \Rightarrow N(0,1). \tag{4.47}$$

III. In the general case we may assume $g(0) \neq 0$, by Theorem 3.14. Then

$$c_\epsilon = g(0)n_\epsilon + (c_\epsilon - g(0)n_\epsilon).$$

By Theorem 3.14 (with a slight modification if the integrand of m_1 below is not asymptotically monotone),

$$\sqrt{\frac{\epsilon}{v_1}}\Big((c_\epsilon - g(0)n_\epsilon - \frac{m_1}{\epsilon}\Big) \underset{\epsilon \to 0}{\Rightarrow} N(0,1),$$

where

$$m_1 = \int_0^\infty \frac{(g(E) - g(0))}{e^{\beta E} - 1} d\mu; \qquad v_1 = \int_0^\infty \frac{(g(E) - g(0))^2 e^{\beta E}}{(e^{\beta E} - 1)^2} d\mu.$$

Together with the result of Part II, this gives

$$\sqrt{\epsilon}\Big(c_\epsilon - \frac{m}{\epsilon}\Big) \Rightarrow N\Big(0, \frac{g(0)a}{\beta^2}\Big),$$

completing the proof. ∎

CHAPTER 5

SINGULAR CENTRAL LIMIT THEOREMS, II

This chapter deals with the most singular families of geometric r.v.'s arising under our hypotheses, those from spectral measures with non-vanishing density near 0. The singularity of this situation fundamentally changes both the results and the proof of the associated limit theorem. The normal distributions of integrals appearing in previous cases are now replaced by an extreme value distribution whose parameters are entirely determined by spectral density at 0. This will be shown in the prototypical case $d\mu(E) = dF(E) = dE$, and in general through an approach like that of Chapter 4. The approximately linear behavior of $F(E)$ at 0 will be exploited to decompose c_ϵ into two independent r.v.'s, to the first of which the prototypical case will apply, and to the second Theorem 4.3.

The *extreme value distribution* is that with d.f. $e^{-e^{-x}}$. The normalized distribution, with zero mean and unit variance, has d.f.

$$\exp\left(-e^{-\left(\frac{\pi}{\sqrt{6}}x+\gamma\right)}\right), \tag{5.1}$$

where $\gamma = .577\ldots$ is Euler's constant. For applications of this distribution to the statistics of extremes, see [G].

§5.1 Asymptotics Under Uniform Spectral Density

In this section c_ϵ, μ, and g are defined as before. We will require the following simple lemma.

LEMMA 5.1: *Let* $\{F_\epsilon(\cdot)\}_{\epsilon>0}$ *be a net of probability d.f.'s, and*

$$F_\epsilon(x) \underset{\epsilon\to 0}{\to} F(x) \qquad (x \in \mathbf{R}), \tag{5.2}$$

where $F(\cdot)$ *is a continuous d.f. If* $\{G_\epsilon(\cdot)\}_{\epsilon>0}$ *is any net of d.f.'s, then*

$$\limsup_{\epsilon\to 0} F_\epsilon(x) * G_\epsilon(x) = \limsup_{\epsilon\to 0} F(x) * G_\epsilon(x) \qquad (x \in \mathbf{R}); \tag{5.3}$$

(5.3) holds as well for the lim inf.

LEMMA 5.2: *If* $F(E) = bE$ *for some* $b > 0$, *and* $g(E) \equiv 1$, *then*

$$R_\epsilon \equiv \epsilon\left(c_\epsilon - \frac{b\left|\ln\left(\frac{\beta\epsilon}{b}\right)\right|}{\epsilon}\right) \underset{\epsilon\to 0}{\Rightarrow} e^{-e^{-\frac{\beta x}{b}}}.$$

Proof: We may assume $b = 1$. The logarithm of the ch.f. of R_ϵ is

$$\ln \Phi_{R_\epsilon}(t) = \ln\left(\mathcal{E}\left(e^{it\left(\epsilon R_\epsilon + \frac{\ln(\beta\epsilon)}{\beta}\right)}\right)\right) = it\frac{\ln(\beta\epsilon)}{\beta} + \sum_{j=1}^{\infty} \ln\left(\frac{1-e^{-\beta\epsilon j}}{1-e^{-\beta\epsilon j + i\epsilon t}}\right), \tag{5.4}$$

where the branch of each term is principal for $t > 0$, and determined by analytic continuation from $t > 0$ elsewhere. Fixing t,

$$\begin{aligned} \ln \Phi_{R_\epsilon}(t) = {} & \frac{it\ln(\beta\epsilon)}{\beta} \\ & + \sum_{j=1}^{\left[\frac{1}{\beta\epsilon}\right]} \left(\ln(1-e^{-\beta\epsilon j}) - \ln(\beta\epsilon j) - \ln(1-e^{-\beta\epsilon j + it\epsilon}) + \ln(\beta\epsilon j - it\epsilon)\right) \\ & + \sum_{j=1}^{\left[\frac{1}{\beta\epsilon}\right]} \left(\ln(\beta\epsilon j) - \ln(\beta\epsilon j - it\epsilon)\right) \\ & + \sum_{j=\left[\frac{1}{\beta\epsilon}\right]+1} \left(\ln(1-e^{-\beta\epsilon j}) - \ln\left(1-e^{(-\beta\epsilon j + i\epsilon t)}\right)\right). \end{aligned} \tag{5.5}$$

Let

$$h(x) = \ln\left(\frac{1-e^{-x}}{x}\right), \qquad l(x) = \ln(1-e^{-x}),$$

with branches as before. The function $h(\cdot)$ is analytic at all points within unit distance of the real axis interval $[0,1]$, and $l(\cdot)$ is analytic at all points within distance $\frac{7}{8}$ of $[0,\infty)$. Denote these regions of analyticity by D_h and D_l, respectively. Let $t \neq 0$ and $|\epsilon t| < \frac{1}{2}$. If $x \in [0,\infty)$, then

$$l(x - i\epsilon t) = l(x) - i\epsilon t l'(x) + R(x),$$

where

$$\begin{aligned} \left|\frac{R(x)}{(i\epsilon t)^2 l'(x)}\right| &= \left|\frac{\ln\left(\frac{1-e^{-(x-i\epsilon t)}}{1-e^{-x}}\right)}{\left(\frac{(\epsilon t)^2}{e^x-1}\right)} - \frac{1}{i\epsilon t}\right| = \left|\frac{\left(\frac{1-e^{i\epsilon t}}{e^x-1}\right) + O\left(\left(\frac{e^{-i\epsilon t}-1}{e^x-1}\right)^2\right)}{\left(\frac{(\epsilon t)^2}{e^x-1}\right)} - \frac{1}{i\epsilon t}\right| \\ &\le \frac{O\left(\left(\frac{e^{-i\epsilon t}-1}{e^x-1}\right)\right)^2}{\left(\frac{(\epsilon t)^2}{e^x-1}\right)} + O(1). \end{aligned} \tag{5.6}$$

The term $O\left(\frac{e^{-i\epsilon t}-1}{e^x-1}\right)$ satisfies

$$\frac{O\left(\frac{e^{-i\epsilon t}-1}{e^x-1}\right)}{\left(\frac{e^{-i\epsilon t}-1}{e^x-1}\right)} \le C,$$

for some $C > 0$ and sufficiently small values of the argument, and

$$O(1) = \frac{e^{-i\epsilon t} - 1}{(\epsilon t)^2} - \frac{1}{\epsilon t} \tag{5.7}$$

remains bounded for all $\epsilon > 0$, uniformly in $x \in D_l$ (in fact independently of x). Thus if $t \neq 0$ is fixed, $\frac{R(x)}{(i\epsilon t)^2 l'(x)}$ is bounded uniformly in x for $x \in D_l$ and ϵ sufficiently small, and

$$\begin{aligned} &\sum_{j=[\frac{1}{\beta\epsilon}]+1}^{\infty} \left(\ln(1 - e^{-\beta\epsilon j}) - \ln(1 - e^{-\beta\epsilon j + i\epsilon t})\right) \\ &= \left\{\sum_{j=[\frac{1}{\beta\epsilon}]+1}^{\infty} i\epsilon t l'(\beta\epsilon j)\right\}(1 + O(\epsilon)) = \left\{\sum_{j=[\frac{1}{\beta\epsilon}]+1}^{\infty} \frac{i\epsilon t}{e^{\beta\epsilon j} - 1}\right\}(1 + O(\epsilon)) \quad (\epsilon \to 0), \end{aligned} \tag{5.8}$$

which clearly also holds if $t = 0$.

Similarly, one can show that for $t \in \mathbf{R}$,

$$\begin{aligned} &\sum_{j=1}^{[\frac{1}{\beta\epsilon}]} \left(\ln(1 - e^{-\beta\epsilon j}) - \ln(\beta\epsilon j) - \ln(1 - e^{-\beta\epsilon j + i\epsilon t}) + \ln(\beta\epsilon j - i\epsilon t)\right) \\ &= \left\{i\epsilon t \sum_{j=1}^{[\frac{1}{\beta\epsilon}]} \left(\frac{1}{e^{\beta\epsilon j} - 1} - \frac{1}{\beta\epsilon j}\right)\right\}(1 + O(\epsilon)) \qquad (\epsilon \to 0). \end{aligned} \tag{5.9}$$

The summands in the last terms of (5.8) and (5.9) are decreasing in $\beta\epsilon j$, and Proposition 3.13 implies

$$\sum_{j=[\frac{1}{\beta\epsilon}]+1}^{\infty} \frac{i\epsilon t}{e^{\beta\epsilon j} - 1} = -\frac{it}{\beta} \ln(1 - e^{-1}) + o(1) \tag{5.10}$$

and

$$i\epsilon t \sum_{j=1}^{[\frac{1}{\beta\epsilon}]} \left(\frac{1}{e^{\beta\epsilon j} - 1} - \frac{1}{\beta\epsilon j}\right) = \frac{it}{\beta} \ln(1 - e^{-1}) + o(1) \qquad (\epsilon \to 0). \tag{5.11}$$

Finally,

$$\begin{aligned} \sum_{j=1}^{[\frac{1}{\beta\epsilon}]} (\ln(\beta\epsilon j)) - \ln(\beta\epsilon j - i\epsilon t) &= \ln\left(\prod_{j=1}^{[\frac{1}{\beta\epsilon}]} j\right) - \ln\left(\prod_{j=1}^{[\frac{1}{\beta\epsilon}]} \left(j - \frac{it}{\beta}\right)\right) \\ &= \ln\Gamma\left(\left[\frac{1}{\beta\epsilon}\right] + 1\right) - \ln\Gamma\left(\left[\frac{1}{\beta\epsilon}\right] - \frac{it}{\beta} + 1\right) + \ln\Gamma\left(1 - \frac{it}{\beta}\right), \end{aligned} \tag{5.12}$$

where we continue our convention on branches. Stirling's expansion gives

$$\ln\Gamma(z) = \left(z - \frac{1}{2}\right)\ln z - z + \frac{1}{2}\ln(2\pi) + O\left(\frac{1}{|z|}\right), \tag{5.13}$$

uniformly for $\arg z \leq \Delta < \pi$. Thus

$$\begin{aligned}
&\ln\Gamma\left(1 - \frac{it}{\beta}\right) + \ln\Gamma\left(\left[\frac{1}{\beta\epsilon}\right] + 1\right) - \ln\Gamma\left(\left[\frac{1}{\beta\epsilon}\right] - \frac{it}{\beta} + 1\right) \\
&= \ln\Gamma\left(1 - \frac{it}{\beta}\right) + \left(\left[\frac{1}{\beta\epsilon}\right] + \frac{1}{2}\right)\left\{\ln\left(\left[\frac{1}{\beta\epsilon}\right] + 1\right) - \ln\left(\left[\frac{1}{\beta\epsilon}\right] - \frac{it}{\beta} + 1\right)\right\} \\
&\qquad + \frac{it}{\beta}\ln\left(\left[\frac{1}{\beta\epsilon}\right] - \frac{it}{\beta} + 1\right) - \frac{it}{\beta} + O(\epsilon) \\
&= \ln\Gamma\left(1 - \frac{it}{\beta}\right) - \left(\left[\frac{1}{\beta\epsilon}\right] + \frac{1}{2}\right)\left(-\frac{it}{\beta}\left(\left[\frac{1}{\beta\epsilon}\right] + 1\right)^{-1} + O(\epsilon^2)\right) \\
&\qquad - \frac{it}{\beta}\ln(\beta\epsilon) - \frac{it}{\beta} + O(\epsilon) \\
&= \ln\Gamma\left(1 - \frac{it}{\beta}\right) - \frac{it}{\beta}\ln(\beta\epsilon) + O(\epsilon) \qquad (\epsilon \to 0).
\end{aligned} \tag{5.14}$$

Combining (5.5) and (5.8-14),

$$\ln(\Phi_{R_\epsilon}(t)) = \ln\Gamma\left(1 - \frac{it}{\beta}\right) + o(1),$$

so that

$$\Phi_{R_\epsilon}(t) = \Gamma\left(1 - \frac{it}{\beta}\right)(1 + o(1)) \underset{\epsilon\to 0}{\to} \Gamma\left(1 - \frac{it}{\beta}\right). \tag{5.15}$$

The right side is the Fourier transform of the distribution with d.f. $e^{-e^{-\beta x}}$, and the proof is completed by the Lévy-Cramér convergence theorem.∎

§5.2 Asymptotics for Non-Vanishing Densities Near Zero

THEOREM 5.3: *If* $\mu \in S_{g,\beta}$, $F'(0) = b > 0$, *and* $g(0) > 0$, *then*

$$R_\epsilon \equiv \epsilon\left(c_\epsilon - \frac{m(\epsilon)}{\epsilon}\right) \underset{\epsilon\to 0}{\Rightarrow} e^{-e^{-\frac{\beta x}{g(0)b}}}$$

where

$$m(\epsilon) = m + \frac{g(0)b}{\beta}\left|\ln\left(\frac{\beta\epsilon}{b}\right)\right|,$$

$$m = \int_0^\infty \frac{g(0)}{e^{\beta E}-1} d(F(E) - bE) + \int_0^\infty \frac{g(E)-g(0)}{e^{\beta E}-1} dF(E). \tag{5.16}$$

Proof: I. We assume initially that $g(E) \equiv 1$ and

$$F''(E) < 0 \qquad (0 \le E \le \lambda) \tag{5.17}$$

for a $\lambda > 0$. Let $F''(0) = -a$, $F^\lambda(E)$ be as in 4.13, and

$$H_\epsilon = \left[\frac{F^\lambda(E)}{\epsilon}\right] + \left[\frac{bE - F^\lambda(E)}{\epsilon}\right]. \tag{5.18}$$

Let $\mathcal{F}_\epsilon^* = \{E_{\epsilon j}^*\}_{1 \le j < \infty}$ be as in Theorem 4.3, and $\mathcal{F}_\epsilon^b = \{E_{\epsilon j}^b\}_{1 \le j < \infty}$ be the discontinuities of $\left[\frac{bE}{\epsilon}\right]$, i.e., $E_{\epsilon j}^b = \frac{\epsilon j}{b}$.

We have

$$0 \le \left[\frac{bE}{\epsilon}\right] - H_\epsilon(E) \le 1 \qquad (E \in \mathbf{R}^+) \tag{5.19}$$

and

$$E_{\epsilon j}^b \le E_{\epsilon j}^* \quad (1 \le j < \infty), \qquad E_{\epsilon(j-1)}^* \le E_{\epsilon j}^b \quad (2 \le j < \infty). \tag{5.20}$$

Let $\{n_{\epsilon j}^*\}_{1 \le j < \infty}$ and $\{n_{\epsilon j}^b\}_{1 \le j < \infty}$, n_ϵ^* and n_ϵ^b correspond as before to $\mathcal{F}_\epsilon^*$ and $\mathcal{F}_\epsilon^b$ (see 4.19). By the lemma,

$$Q_\epsilon^b \equiv \left(n_\epsilon^b + \frac{b\ln\left(\frac{\beta\epsilon}{b}\right)}{\beta\epsilon}\right) \underset{\epsilon\to 0}{\Rightarrow} e^{-e^{-\frac{\beta}{b}x}} \tag{5.21}$$

Let

$$n_\epsilon^{b,j} = n_\epsilon^b - n_{\epsilon j}^b; \qquad n_\epsilon^{*,j} = n_\epsilon^* - n_{\epsilon j}^*;$$

$$n_{\epsilon \check{j}}^b = \sum_{k=1}^{j} n_{\epsilon j}^b; \qquad n_{\epsilon \check{j}}^* = \sum_{k=1}^{j} n_{\epsilon j}^* \qquad (\epsilon > 0, \quad j = 2, 3, 4 \ldots). \tag{5.22}$$

The d.f. of $\epsilon n_{\epsilon j}^b$ is

$$F_{n_{\epsilon j}^b}(x) = 1 - e^{\beta\left[\frac{x}{\epsilon}+1\right]\frac{\epsilon j}{b}} \underset{\epsilon\to 0}{\rightarrow} 1 - e^{-\frac{\beta j}{b}x} \tag{5.23}$$

and thus

$$W_{\epsilon j}^b \equiv \epsilon n_{\epsilon j}^b \underset{\epsilon\to 0}{\Rightarrow} 1 - e^{-\frac{\beta j}{b}x}. \tag{5.24}$$

We subdivide $\mathcal{F}_\epsilon^*$ as in Theorem 4.3 into $\mathcal{F}_\epsilon^{*I}$ and $\mathcal{F}_\epsilon^{*II}$ with S_ϵ^I and S_ϵ^{II} defined as before. Since

$$\frac{bE - F^\lambda(E)}{F^\lambda(E)} \underset{E\to 0}{\to} 0,$$

for $j \in \mathrm{N}$ there is an $\epsilon > 0$ such that if $\epsilon' \leq \epsilon$, then

$$\left[\frac{bE_{\epsilon' j}^{*I} - F^\lambda(E_{\epsilon' j}^{*I})}{\epsilon'}\right] = 0,$$

since

$$E_{\epsilon j}^{*I} = \inf\left\{E' : \frac{F^\lambda(E')}{\epsilon} \geq j\right\}.$$

Thus if ϵ is sufficiently small and k is fixed

$$E_{\epsilon j}^* \in \mathcal{F}_\epsilon^{*I} \qquad (j \leq k \in \mathrm{N}). \tag{5.25}$$

Since $F'(0) = b$ and by (5.17), there exist $E_M > 0$ and $L > 0$ such that if $E \leq E_M$,

$$bE - LE^2 \leq F^\lambda(E) \leq bE. \tag{5.26}$$

Let $j \in \mathrm{N}$, and ϵ be sufficiently small that (i) equation (5.25) holds for some $k \geq j$ and (ii) $E_{\epsilon j}^* \leq E_M$. Then

$$E_{\epsilon j}^* = E_{\epsilon j}^{*I} = \inf\left\{E' : \frac{F^\lambda(E')}{\epsilon} \geq j\right\}.$$

Together with (5.26) this implies

$$E_{\epsilon j}^* = \frac{j\epsilon}{b} + O(\epsilon^2) \qquad (\epsilon \to 0). \tag{5.27}$$

Thus, as in (5.23),

$$F_{\epsilon n_{\epsilon j}^*}(x) = 1 - e^{-\beta[\frac{x}{\epsilon}+1](\frac{j\epsilon}{b}+O(\epsilon^2))} \underset{\epsilon\to 0}{\to} 1 - e^{-\frac{\beta j}{b}x}. \tag{5.28}$$

Let $\mathcal{D}(\alpha)$ denote the exponential distribution with distribution function $1 - e^{-\alpha x}$ and

$$\mathcal{D}^j \equiv \mathcal{D}\left(\frac{\beta}{b}\right) * \mathcal{D}\left(\frac{2\beta}{b}\right) * \cdots * \mathcal{D}\left(\frac{j\beta}{b}\right).$$

Then by (5.24) and (5.28)

$$W_{\epsilon \check{j}}^b \equiv \epsilon n_{\epsilon \check{j}}^b \underset{\epsilon\to 0}{\Rightarrow} \mathcal{D}^j; \qquad W_{\epsilon \check{j}}^* \equiv \epsilon n_{\epsilon \check{j}}^* \underset{\epsilon\to 0}{\Rightarrow} \mathcal{D}^j \qquad (j \in \mathrm{N}). \tag{5.29}$$

Let

$$Q_\epsilon^{b,j} = \epsilon\left(n_\epsilon^{b,j} - \frac{m(\epsilon)}{\epsilon}\right),$$

$$n_\epsilon^{b,\check{j}} = n_\epsilon^b - n_{\epsilon\check{j}}^b, \qquad n_\epsilon^{*,\check{j}} = n_\epsilon^* - n_{\epsilon\check{j}}^*,$$

and

$$Q_\epsilon^{b,\check{j}} = \epsilon\left(n_\epsilon^{b,\check{j}} - \frac{m(\epsilon)}{\epsilon}\right), \qquad Q_\epsilon^{*,\check{j}} = \epsilon\left(n_\epsilon^{*,\check{j}} - \frac{m(\epsilon)}{\epsilon}\right). \tag{5.30}$$

By (5.20)

$$F_{n^*_{\epsilon(j-1)}}(x) \le F_{n^b_{\epsilon j}}(x) \qquad (\epsilon > 0,\ x \in \mathbf{R},\ j = 2, 3, \dots\ ),$$

and hence

$$F_{n_\epsilon^{*,\check{j}}}(x) \le F_{n_\epsilon^{b,k}}(x) \qquad (k = j+1), \tag{5.31}$$

so that

$$F_{Q_\epsilon^{*,\check{j}}}(x) \le F_{Q_\epsilon^{b,k}}(x) \qquad (k = j+1). \tag{5.32}$$

Let

$$Q_\epsilon^* = \epsilon\left(n_\epsilon^* - \frac{b|\ln(\frac{\beta\epsilon}{b})|}{\beta\epsilon}\right). \tag{5.33}$$

Thus by Lemma 5.1

$$\begin{aligned}
\limsup_{\epsilon\to 0} F_{Q_\epsilon^*}(x) &= \limsup_{\epsilon\to 0} F_{W^*_{\epsilon\check{j}}} * F_{Q_\epsilon^{*,\check{j}}}(x) = \limsup_{\epsilon\to 0} F_{D^j} * F_{Q_\epsilon^{*,\check{j}}}(x) \\
&\le \limsup_{\epsilon\to 0} F_{D^j} * F_{Q_\epsilon^{b,k}}(x) = \limsup_{\epsilon\to 0} F_{W_\epsilon^{b,\check{j}}} * F_{Q_\epsilon^{b,k}}(x)\ , \\
&= \limsup_{\epsilon\to 0} F_{Q_\epsilon{}^{b,k}}(x) \qquad (x \in \mathbf{R},\ j \in \mathbf{N},\ k = j+1),
\end{aligned} \tag{5.34}$$

where we have also used (5.29), (5.31), (5.32), Lemma 4.2, and (5.30).

We have $Q_\epsilon^{b,j} * W_{\epsilon j}^b = Q_\epsilon^b$, and thus

$$\Phi_{Q_\epsilon^{b,\check{j}}}(t)\Phi_{W^b_{\epsilon j}}(t) = \Phi_{Q^b_\epsilon}(t) \qquad (\epsilon > 0,\ t \in \mathbf{R},\ j \in \mathbf{N}). \tag{5.35}$$

By (5.21) and (5.24),

$$\Phi_{Q^b_\epsilon}(t) \underset{\epsilon\to 0}{\to} \Gamma\left(1 - \frac{bit}{\beta}\right); \qquad \Phi_{W^b_{\epsilon j}}(t) \underset{\epsilon\to 0}{\to} \left(1 - \frac{bit}{\beta}\right)^{-1} \qquad (t \in \mathbf{R},\ j \in \mathbf{N}),$$

the right sides being characteristic functions of $e^{-e^{-\frac{\beta}{b}x}}$ and $\mathcal{D}\left(\frac{\beta}{b}j\right)$, respectively. Thus

$$\Phi_{Q_\epsilon^{b,j}}(t) \underset{\epsilon\to 0}{\to} \left(1 - \frac{bit}{\beta j}\right)\Gamma\left(1 - \frac{bit}{\beta}\right) \equiv \Phi_b^j(t). \tag{5.36}$$

Since $\Phi_b^j(\cdot)$ is continuous at 0,

$$F_{Q_\epsilon^{b,j}} \underset{\epsilon \to 0}{\to} F_b^j(x)$$

where

$$\int_{-\infty}^{\infty} e^{ixt} dF_b^j(x) = \Phi_b^j(t),$$

and

$$\limsup_{\epsilon \to 0} F_{Q_\epsilon^{b,j}}(x) \leq F_b^j(x) \qquad (x \in \mathbf{R},\ j \in \mathbf{N}). \tag{5.37}$$

Thus by (5.34) and (5.37)

$$\limsup_{\epsilon \to 0} F_{Q_\epsilon^*}(x) \leq F_b^j(x) \qquad (j \in \mathbf{N}). \tag{5.38}$$

On the other hand, by (5.3.6),

$$F_b^j(x) \underset{j \to \infty}{\to} e^{-e^{-\frac{\beta}{b}x}}.$$

Since the left side of (5.38) is independent of j,

$$\limsup_{\epsilon \to 0} F_{Q_\epsilon^*}(x) \leq e^{-e^{-\frac{\beta}{b}x}}. \tag{5.39}$$

By (5.20),

$$F_{n_{\epsilon j}^b}(x) \leq F_{n_{\epsilon j}^*}(x) \qquad (\epsilon > 0,\ x \in \mathbf{R},\ j \in \mathbf{N}), \tag{5.40}$$

so that

$$F_{Q_\epsilon^b}(x) \leq F_{Q_\epsilon^*}(x). \tag{5.41}$$

Combining (5.21), (5.41) and (5.39) yields

$$e^{-e^{-\frac{\beta}{b}x}} = \lim_{\epsilon \to 0} F_{Q_\epsilon^b}(x) \leq \liminf_{\epsilon \to 0} F_{Q_\epsilon^*}(x) \leq \limsup_{\epsilon \to 0} F_{Q_\epsilon^*}(x) \leq e^{-e^{-\frac{\beta}{b}x}}$$

so that

$$F_{Q_\epsilon^*}(x) \underset{\epsilon \to 0}{\to} e^{-e^{-\frac{\beta}{b}x}} \qquad (x \in \mathbf{R}) \tag{5.42}$$

and

$$Q_\epsilon^* \Rightarrow e^{-e^{-\frac{\beta}{b}x}}. \tag{5.43}$$

We now form

$$n_\epsilon^* = n_\epsilon^{*I} + n_\epsilon^{*II}, \tag{5.44}$$

where n_ϵ^{*I} and n_ϵ^{*II} are defined as in (4.25). By arguments identical to those following equation (4.26), n_ϵ^{*I} is identically distributed with n_ϵ^λ, the occupation number corresponding to $\frac{F^\lambda}{\epsilon}$. We have analogously (by Theorem 4.3)

$$Q_\epsilon^{*II} = \frac{\phi(\epsilon)\beta}{\sqrt{a}}\left(n_\epsilon^{*II} - \frac{m^{*II}}{\epsilon}\right) \underset{\epsilon\to 0}{\Rightarrow} N(0,1), \tag{5.45}$$

where

$$m^{*II} = \int_0^\infty \frac{1}{e^{\beta E} - 1} d(bE - F^\lambda(E)),$$

$$\phi(\epsilon) = \sqrt{\frac{\epsilon}{|\ln \epsilon|}}. \tag{5.46}$$

By (5.33),

$$Q_\epsilon^* = \epsilon\left(n_\epsilon^{*I} - \left(\frac{b}{\epsilon\beta}\left|\ln\left(\frac{\beta\epsilon}{b}\right)\right| - \frac{m^{*II}}{\epsilon}\right)\right) + \frac{\sqrt{a}\epsilon}{\beta\phi(\epsilon)} Q_\epsilon^{*II}. \tag{5.47}$$

Hence by (5.45) and (5.43),

$$\epsilon\left(n_\epsilon^\lambda - \left(\frac{b}{\epsilon\beta}\left|\ln\left(\frac{\beta}{b}\epsilon\right)\right| - \frac{m^{*II}}{\epsilon}\right)\right) \underset{\epsilon\to 0}{\Rightarrow} e^{-e^{-\frac{\beta}{b}x}}. \tag{5.48}$$

Let

$$n_\epsilon^>(\lambda) \equiv n_\epsilon - n_\epsilon^\lambda.$$

As in (4.33),

$$\sqrt{\frac{\epsilon}{v^>(\lambda)}}\left(n_\epsilon^>(\lambda) - \frac{m^>(\lambda)}{\epsilon}\right) \underset{\epsilon\to 0}{\Rightarrow} N(0,1), \tag{5.49}$$

with $m^>(\lambda)$ and $v^>(\lambda)$ defined as before. Thus,

$$\begin{aligned}\epsilon\left(n_\epsilon - \frac{m(\epsilon)}{\epsilon}\right) &= \epsilon\left(n_\epsilon^\lambda(\lambda) - \left(\frac{b}{\epsilon\beta}\left|\ln\left(\frac{\beta}{b}\epsilon\right)\right| - \frac{m^{*II}}{\epsilon}\right)\right) \\ &\quad + \epsilon\left(n^>(\lambda) - \frac{m^{*II}}{\epsilon} + \frac{b}{\epsilon\beta}\left|\ln\left(\frac{\beta}{b}\epsilon\right)\right| - \frac{m(\epsilon)}{\epsilon}\right).\end{aligned} \tag{5.50}$$

By calculation,

$$m^{*II} - \frac{b}{\epsilon\beta}\left|\ln\left(\frac{\beta\epsilon}{b}\right)\right| - m(\epsilon) = m^>(\lambda); \tag{5.51}$$

hence the second term in (5.50) converges to δ_0. By (5.48) and (5.50),

$$R_\epsilon = \frac{1}{\sigma(\epsilon)}\left(n_\epsilon - \frac{m(\epsilon)}{\epsilon}\right) \underset{\epsilon\to 0}{\Rightarrow} e^{-e^{-\frac{\beta x}{b}}}. \tag{5.52}$$

II. We eliminate here the restriction on $F''(0)$. As in Theorem 4.3, let $\mu^0 \in S_{1,\beta}$ have spectral function $F^0(\cdot)$ such that $(i) F(E) - F^0(E)$ is monotone increasing in E, $(ii) (F^0)'(0) =$

b, and (iii) $(F^0)''(0) = -a < 0$. Let $F^*(\cdot)$ be defined by (4.37). We employ the definitions between (4.37) and (4.40). Equation (4.38) holds here also, and yields

$$F_{R_\epsilon}(x) \le F_{R_\epsilon^*}(x) \qquad (x \in \mathbf{R},\ \epsilon > 0), \tag{5.53}$$

where

$$R_\epsilon^* = \epsilon\left(n_\epsilon^* - \frac{m(\epsilon)}{\epsilon}\right), \qquad R_\epsilon = \epsilon\left(n_\epsilon - \frac{m(\epsilon)}{\epsilon}\right).$$

We use arguments like those in (5.26-27) to conclude

$$E_{\epsilon j} = \frac{j\epsilon}{b} + O(\epsilon^2), \qquad E_{\epsilon j}^* = \frac{j\epsilon}{b} + O(\epsilon^2).$$

Thus, as in (5.27-29),

$$W_{\epsilon\check{j}} \equiv \epsilon n_{\epsilon\check{j}} \underset{\epsilon\to 0}{\Rightarrow} \mathcal{D}^j, \qquad W_{\epsilon\check{j}}^* \equiv \epsilon n_{\epsilon\check{j}}^* \underset{\epsilon\to 0}{\Rightarrow} \mathcal{D}^j, \tag{5.54}$$

where $n_{\epsilon\check{j}}$ and $n_{\epsilon\check{j}}^*$ are defined as in (5.22). As in (5.34),

$$\liminf_{\epsilon\to 0} F_{R_\epsilon^*}(x) \le \liminf_{\epsilon\to 0} F_{R_\epsilon^j}(x), \tag{5.55}$$

where

$$R_\epsilon^j = \epsilon\left(n_\epsilon^j - \frac{m(\epsilon)}{\epsilon}\right), \qquad n_\epsilon^j = n_\epsilon - n_{\epsilon j}.$$

In (5.55), we have used Lemma 5.1, and

$$F_{R_\epsilon^{*\check{j}}}(x) \le F_{R_\epsilon^k}(x) \qquad (j \in \mathbf{N},\ k = j+1), \tag{5.56}$$

where

$$R_\epsilon^{\check{j}} = \frac{1}{\sigma(\epsilon)}\left(n_\epsilon - n_{\epsilon\check{j}} - \frac{m(\epsilon)}{\epsilon}\right)$$

and $R_\epsilon^{*\check{j}}$ is defined similarly. By Lemma 5.1 and (5.55)

$$\liminf_{\epsilon\to 0} F_{R_\epsilon^*} * F_{W_{\epsilon j}}(x) \le \liminf_{\epsilon\to 0} F_{R_\epsilon^j} * F_{W_{\epsilon j}}(x) = \liminf_{\epsilon\to 0} F_{R_\epsilon}(x) \tag{5.57}$$

where

$$W_{\epsilon j} \equiv \epsilon n_{\epsilon j} \qquad (\epsilon > 0,\ j \in \mathbf{N}).$$

Define n_ϵ^{*I} and n_ϵ^{*II} as in (4.25), with respect to the decomposition

$$\mathcal{F}_\epsilon^* = \mathcal{F}_\epsilon^{*I} \cup \mathcal{F}_\epsilon^{*II},$$

where $\mathcal{F}_\epsilon^{*I}$ corresponds to the discontinuities of $\left[\frac{F^0(E)}{\epsilon}\right]$ and $\mathcal{F}_\epsilon^{*II}$ to those of $\left[\frac{F(E)-F^0(E)}{\epsilon}\right]$. By Theorem 4.3,

$$\frac{\phi(\epsilon)\beta}{\sqrt{a}}\left(n_\epsilon^{*II}-\frac{m^{*II}}{\epsilon}\right)\underset{\epsilon\to 0}{\Rightarrow} N(0,1), \tag{5.58}$$

with m^{*II} as in (4.42). The measure μ^0 satisfies the hypotheses of Part I, so that

$$\epsilon\left(n_\epsilon^{*I}-\frac{m^{*I}(\epsilon)}{\epsilon}\right)\underset{\epsilon\to 0}{\Rightarrow} e^{-e^{-\frac{\beta x}{b}}}, \tag{5.59}$$

where

$$m^{*I}(\epsilon)=m^{*I}+\frac{b\left|\ln\left(\frac{\beta\epsilon}{b}\right)\right|}{\beta},\qquad m^{*I}=\int_0^\infty\frac{1}{e^{\beta E}-1}d(F^0(E)-bE). \tag{5.60}$$

Combining (5.58) and (5.59) yields

$$R_\epsilon^*=\epsilon\left(n_\epsilon^{*I}-\frac{m^{*I}(\epsilon)}{\epsilon}\right)+\epsilon\left(n_\epsilon^{*II}-\frac{m(\epsilon)}{\epsilon}+m^{*I}(\epsilon)\right). \tag{5.61}$$

We have

$$m(\epsilon)-m^{*I}(\epsilon)=m^{*II}$$

so that by (5.58), the second term on the right of (5.61) converges to δ_0, and

$$R_\epsilon^*\underset{\epsilon\to 0}{\Rightarrow} e^{-e^{-\frac{\beta x}{b}}}. \tag{5.62}$$

By Lemma 5.1, (5.57), (5.53), and (5.62),

$$\begin{aligned} e^{-e^{-\frac{\beta x}{b}}}*F_{D(\frac{\beta j}{b})}&=\liminf_{\epsilon\to 0}F_{R_\epsilon^*}*F_{W_{\epsilon j}}(x)\le\liminf_{\epsilon\to 0}F_{R_\epsilon}(x)\le\limsup_{\epsilon\to 0}F_{R_\epsilon}(x)\\ &\le\lim_{\epsilon\to 0}F_{R_\epsilon^*}(x)=e^{-e^{-\frac{\beta x}{b}}}. \end{aligned} \tag{5.63}$$

The limit $j\to\infty$ yields

$$F_{D(\frac{\beta j}{b})}(x)\underset{j\to\infty}{\to}F_{\delta_0}(x) \tag{5.64}$$

so that

$$R_\epsilon\Rightarrow e^{-e^{-\frac{\beta x}{b}}}.$$

III. We finally consider general c_ϵ for which g is not identically 1. We have

$$c_\epsilon=g(0)n_\epsilon+(c_\epsilon-g(0)n_\epsilon).$$

By Theorem 3.1,

$$\phi(\epsilon)\Big(c_\epsilon - g(0)n_\epsilon - \frac{m_1}{\epsilon}\Big) \underset{\epsilon\to 0}{\Rightarrow} N(0, v_1), \tag{5.65}$$

with

$$m_1 = \int_0^\infty \frac{g(E) - g(0)}{e^{\beta E} - 1} d\mu, \qquad v_1 = \int_0^\infty \frac{(g(E) - g(0))^2 e^{\beta E}}{(e^{\beta E} - 1)^2} d\mu.$$

Hence

$$\epsilon\Big(c_\epsilon - \frac{m(\epsilon)}{\epsilon}\Big) = \epsilon\Big((c_\epsilon - g(0)n_\epsilon) - \frac{m_1}{\epsilon}\Big) + \epsilon\Big(g(0)n_\epsilon - \frac{m(\epsilon) - m_1}{\epsilon}\Big) \underset{\epsilon\to 0}{\Rightarrow} e^{-e^{-\frac{\beta x}{g(0)b}}}, \tag{5.66}$$

completing the proof.

CHAPTER 6

PHYSICAL APPLICATIONS

In this chapter we examine particle number and energy distributions in certain canonical ensembles. The means of energy distributions provide generalized Planck laws; the classical "blackbody curve" will be shown to apply in several situations. Normality of observables is however not guaranteed, and the extreme value distribution arises in one-dimensional systems.

To simplify the discussion, we consider only particles with chemical potential 0 (whose energy of creation has infimum 0), e.g., photons and neutrinos. We study these in Minkowski space and Einstein space (a spatial compactification of Minkowski space; see Segal [Se2]), two versions of reference space. The canonical relativistic single particle Hamiltonian in its scalar approximation will be used in both cases.

We study Minkowski space ensembles in arbitrary dimension. The Hamiltonian is continuous, and a net of discrete approximations based on localization in space will be introduced. Einstein space will be considered in its two as well as its (physical) four dimensional version. The approach in these cases can be used to establish photon number and energy distributions in a large class of Riemannian geometries, once the relevant wave equations are solved.

§6.1 The Spectral Measure: an Example in Schrödinger Theory

We begin with remarks about obtaining the spectral measure μ for an operator A with continuous spectrum, acting in Minkowski space. An illustrative example of such an infinite volume limit is in [Si].

Let H be an operator on $L^2(\mathbf{R}^n)$. Let χ_R denote the characteristic function of $B_R = \{x : |x| \leq R\}$ and τ_R be its volume. Suppose that for all $g \in C_c^\infty$

$$\lambda(g) \equiv \lim_{R\to\infty} \tau_R^{-1} \operatorname{Tr}(\chi_R g(H))$$

exists. This defines a positive linear functional on C_c^∞, and there is a Borel measure $d\mu$ defined by

$$\lambda(g) = \int g(\lambda) d\mu(\lambda).$$

This measure is the *density of states*. It is the spectral measure associated with an infinite volume limit, and the term is reserved for situations involving Schrödinger operators.

If H is a Schrödinger operator (relevant in our context for the statistical mechanics of non-relativistic particles), we give a criterion for existence of a density of states [Si]. Assume the dimension n of space is greater than 2, and define the space of potentials

$$K_n = \left\{ V : \mathbf{R}^n \to \mathbf{R} \quad : \lim_{\alpha \to 0} \left[\sup_x \int_{|x-y| \leq \alpha} |x-y|^{-n-2} |V(y)| dy \right] = 0 \right\}.$$

The spaces K_n are more relevant to properties of Schrödinger operators $-\Delta + V$ than are L^p or local L^p spaces. We have

THEOREM 6.1 (see [Si]): *Let $H = -\Delta + V(x)$ act on $\mathbf{R}^n$, with $V \in K_n$. The density of states exists if and only if*

$$\mathcal{L}_n(t) = \lim_{R \to \infty} \tau_R^{-1} \operatorname{Tr}\left(\chi_R e^{-tH}\right)$$

exists.

This density of states is the physically appropriate spectral measure for statistical mechanics.

§6.2. The Spectral Approximation Theorem

Let $g \in G_\beta$, and $\mu \in S_{g,\beta}$ be a spectral measure with spectral function $F(\cdot)$; let A'_ϵ and $n'_{\epsilon j}$ be the ϵ-discrete operator and occupation number r.v.'s corresponding to μ. Let $\{A_\epsilon\}_\epsilon$ be a net of discrete operators with corresponding occupation numbers $n_{\epsilon j}$, and

$$c_\epsilon = \sum_{j=1}^{\infty} g(E_{\epsilon j}) n_{\epsilon j}, \tag{6.1}$$

with c'_ϵ defined similarly. We now ask, How similar must the spectra of A_ϵ and A'_ϵ be in order that c_ϵ and c'_ϵ coincide asymptotically in law?

Let $\mathcal{F}_\epsilon = \{E_{\epsilon j}\}_{j=1}^{\infty}$ be the spectrum of A_ϵ, $N_\epsilon(a,b)$ the cardinality of $\mathcal{F}_\epsilon \cap [a,b]$, and

$$c_\epsilon(a,b) \equiv \sum_{k=1}^{\infty} g(E_{\epsilon j}) \chi(E_{\epsilon j}) n_{\epsilon j}$$

be the truncation of c_ϵ, with $\chi(E)$ the characteristic function of $[a,b]$. We make similar definitions for $\mathcal{F}'_\epsilon, N'_\epsilon$, and c'_ϵ, with respect to A'_ϵ. We will assume that there are positive extended real functions $E_1(\epsilon)$ and $E_2(\epsilon)$ such that

(*i*) $\dfrac{\mathcal{V}(c'_\epsilon(0, E_1(\epsilon)))}{\mathcal{V}(c'_\epsilon)} \underset{\epsilon \to 0}{\longrightarrow} 0$

(*ii*) $\dfrac{\mathcal{V}(c'_\epsilon(0, E_2(\epsilon)))}{\mathcal{V}(c'_\epsilon)} \underset{\epsilon \to 0}{\longrightarrow} 1$

(*iii*) $\sup_{E_1(\epsilon) \leq E \leq E_2(\epsilon)} \left| \dfrac{N_\epsilon(0,E)}{N'_\epsilon(0,E)} \right| \underset{\epsilon \to 0}{\longrightarrow} 0$

(*iv*) $N_\epsilon(0, E) \leq M_1 N'_\epsilon(0, M_2 E) \qquad (0 \leq E \leq E_1(\epsilon))$

for some $M_1, M_2 > 0$, independent of ϵ.

We omit the proof of

THEOREM 6.2: *Under these assumptions, if $F'(0) = 0$ then c_ϵ and c'_ϵ have the same asymptotic law i.e.,*

$$R_\epsilon = \frac{1}{\sqrt{\mathcal{V}(c_\epsilon)}}\left(c'_\epsilon - \frac{m'}{\epsilon}\right) \tag{6.2}$$

and

$$R'_\epsilon = \frac{1}{\sqrt{\mathcal{V}(c_\epsilon)}}\left(c_\epsilon - \frac{m(\epsilon)}{\epsilon}\right)$$

have identical (normal) asymptotic distributions, where

$$m' = \int_0^\infty \frac{g(E)}{e^{\beta E} - 1} d\mu, \tag{6.3}$$

and

$$m(\epsilon) - m' = o(1) \qquad (\epsilon \to 0).$$

COROLLARY 6.2.1: *Conditions (i) and (ii) may be replaced by the generally stronger ones*

(*i'*) $\epsilon \mathcal{V}(c'_\epsilon(0, E_1(\epsilon))) \to 0$

(*ii'*) $\epsilon \mathcal{V}(c'_\epsilon(E_2(\epsilon), \infty)) \to 0$.

§6.3 Observable Distributions in Minkowski Space

We now construct the canonical single particle Hilbert space for $n + 1$ dimensional Minkowski space; the constructions on other spaces are made similarly. All particles will be treated in scalar approximations, so fields will be approximated by scalar fields, and spins by spin 0. Distributions will be studied in the frame in which expected angular and linear momentum vanishes.

We let n dimensions correspond to position

$$(x_1, x_2, \ldots, x_n) = \mathbf{x}$$

and one to time $x_0 = t$. Total space-time coordinates are

$$(x_0, x_1 \ldots, x_n) = \tilde{x};$$

the speed of light is 1 here.

The field $\phi(\cdot)$ of a single Lorentz and translation invariant non-self-interacting scalar particle satisfies

$$\Delta\phi(\tilde{x}) - \phi_{tt}(\tilde{x}) = m^2\phi(\tilde{x}), \tag{6.4}$$

where m is mass, subscripts denote differentiation, and

$$\Delta = \frac{\partial^2}{\partial x_1^2} + \frac{\partial^2}{\partial x_2^2} + \cdots + \frac{\partial^2}{\partial x_n^2}.$$

Since $m = 0$ by assumption, we have $\Delta\phi(\tilde{x}) = \phi_{tt}(\tilde{x})$.

The single particle space $\mathcal{H}$ consisting of solutions of (6.4) is formulated most easily in terms of Fourier transforms. We have $\phi(\cdot) \in \mathcal{H}$ if $\phi(\cdot)$ is real-valued and

$$\phi(\tilde{x}) = (2\pi)^{-\frac{n}{2}} \int_{\mathbf{R}^{n+1}} e^{i\tilde{k}\cdot\tilde{x}} F(\tilde{k}) d\tilde{x}, \tag{6.5}$$

where

$$\tilde{k} = (k_0, k_1, \ldots, k_n), \qquad d\tilde{k} = dk_0 dk_1 \ldots dk_n \tag{6.6}$$

and

$$\tilde{k} \cdot \tilde{x} = k_0 x_0 - \sum_{i=1}^{n} k_i x_i.$$

Above, $F(\cdot)$ is a distribution

$$F(\tilde{k})d\tilde{k} = \{\delta(k_0 - |\mathbf{k}|) f(\mathbf{k}) + \delta(k_0 + |\mathbf{k}|)\overline{f}(-\mathbf{k})\}|\mathbf{k}|^{-1} d\mathbf{k},$$

where

$$\frac{f(\mathbf{k})}{\sqrt{\mathbf{k}}} \in L^2(\mathbf{R}^n); \tag{6.7}$$

$$\mathbf{k} = (k_1, \ldots, k_n), \qquad |\mathbf{k}| = \left(\sum_{i=1}^{n} k_i^2\right)^{\frac{1}{2}}, \qquad d\mathbf{k} = dk_1 dk_2 \ldots dk_n, \tag{6.8}$$

$\delta(\cdot)$ denotes the Dirac delta distribution, and $\overline{f}(\cdot)$ is the complex conjugate of $f(\cdot)$. The inner product of $f_1(\cdot)$, $f_2(\cdot) \in \mathcal{H}$ (we use $\phi(\cdot)$ and $f(\cdot)$ interchangeably) is

$$\langle f_1, f_2 \rangle = \int_{\mathbf{R}^n} \frac{f_1(\mathbf{k}) f_2(\mathbf{k})}{|\mathbf{k}|} d\mathbf{k}. \tag{6.9}$$

Since (6.5) is hyperbolic, solutions correspond to Cauchy data; we have

$$f(\mathbf{k}) = \frac{1}{2(2\pi)^{\frac{n}{2}}} \left\{ |\mathbf{k}| \int_{\mathbf{R}^n} e^{i\mathbf{k}\cdot\mathbf{x}} \phi(0, \mathbf{x}) d\mathbf{x} - i \int_{\mathbf{R}^n} e^{i\mathbf{k}\cdot\mathbf{x}} \phi_t(0, \mathbf{x}) d\mathbf{x} \right\}. \tag{6.10}$$

The state space thus consists of admissible Cauchy data, with time evolution dictated by (6.6). Since solutions $\phi(\cdot)$ are real, the most convenient representation of $\mathcal{H}$, which is complex, is as Lebesgue measurable functions $f(\cdot)$ on $\mathbf{R}^n$ satisfying (6.7), with inner product (6.9). In this representation the Hamiltonian $A = \frac{1}{i}\frac{\partial}{\partial t}$ is multiplication by $|\mathbf{k}|$.

The operator A has continuous spectrum, necessitating a discrete approximation in the formation of a density operator. To this end, we localize in space, replacing $\mathbf{R}^n$ with the torus $\mathbf{T}^n$. Asymptotic distributions are largely independent of the compact manifold used; the sphere will prove to give the same asymptotics.

Let A_ϵ be the single particle Hamiltonian on the Torus $\mathbf{T}^n$ of volume $\frac{(2\pi)^n}{\epsilon}$, with spectrum

$$\mathcal{F}_\epsilon = \left\{ \epsilon \left(\sum_{j=1}^{n} i_j^2 \right)^{\frac{1}{2}} : (i_1, \ldots, i_n) \in (\mathbf{Z}^+)^n \right\} \equiv \{E_{\epsilon j}\}_{j=1}^{\infty} \tag{6.11}$$

(We neglect the 0 eigenvalue, which is without physical consequence.) The particle number and energy r.v.'s corresponding to A_ϵ in its canonical ensemble (at given inverse temperature) will be

$$n_\epsilon(a,b) = \sum_{a \le E_{\epsilon j} \le b} n_{\epsilon j}, \qquad \Delta_\epsilon(a,b) = \sum_{a \le E_{\epsilon j} \le b} E_{\epsilon j} n_{\epsilon j}.$$

DEFINITIONS 6.3: $K_1(\cdot)$ denotes the modified Bessel function of the second kind, of order 1.

Let A'_ϵ be the ϵ-discrete operator corresponding to the spectral measure μ with spectral function

$$F(E) = \frac{\pi^{\frac{n}{2}} E^n}{\Gamma(\frac{n+2}{2})} \qquad (E \ge 0), \tag{6.12}$$

which is the volume of an n-dimensional sphere of radius E.

LEMMA 6.4: *Let $l_n(x)$ be the number of non-zero lattice points within a sphere of radius x in n dimensions. Then*

a) There is a number M_n such that $l_n(x) \le M_n x^n$

b) For each n

$$\frac{l_n(x)\Gamma(\frac{n+2}{2})}{\pi^{\frac{n}{2}} x^n} \to 1 \tag{6.13}$$

Proof: Statement (a) clearly holds for x large and small, and both sides are bounded for intermediate values. Assertion (b) states that the ratio of the volume of a sphere to the number of its lattice points converges to 1.∎

For brevity and convenience we define the following real functions, which are fundamentally related to asymptotic means and variances in canonical ensembles on Riemannian manifolds. They are quantities determined by "global" properties of the appropriate spectral measure μ, and thus arise in those less singular situations in which asymptotics are determined by the totality of μ rather than its behavior near **0**.

DEFINITIONS 6.5: For $n = 0, 1, 2, \dots$ we define

$$m_n^{\pm}(a,b) = \int_a^b \frac{E^{n-1}}{e^{\beta E} \pm 1} dE, \qquad v_n^{\pm}(a,b) = \int_a^b \frac{E^{n-1} e^{\beta E}}{(e^{\beta E} \pm 1)^2} dE.$$

The Riemann zeta function is denoted by $\varsigma(\cdot)$, and

$$\alpha_n = \frac{n\pi^{\frac{n}{2}}}{\Gamma(\frac{n+2}{2})}.$$

In the (important) study of total particle number and energy, we will consider $v_n^{\pm} = v_n^{\pm}(0,\infty)$ and $m_n^{\pm} = m_n^{\pm}(0,\infty)$, which can be calculated explicitly:

$$m_n^- = \frac{\Gamma(n)\varsigma(n)}{\beta^n} \qquad (n = 2, 3, 4, \dots)$$

$$v_n^- = \frac{\Gamma(n)\varsigma(n-1)}{\beta^n} \qquad (n = 3, 4, 5, \dots)$$

$$m_n^+ = \begin{cases} \Gamma(n)\varsigma(n)(1 - 2^{1-n})\beta^{-n}; & (n \geq 2) \\ (\ln 2)\beta^{-1}; & (n = 1) \end{cases}$$

$$v_n^+ = \begin{cases} \Gamma(n)\varsigma(n-1)(1 - 2^{2-n})\beta^{-n}; & (n \geq 3) \\ \beta^{-2}\ln 2; & (n = 2) \\ (2\beta)^{-1}; & (n = 1) \end{cases}$$

THEOREM 6.6: *Under symmetric statistics,*
(1) If $a > 0$ or if $n \geq 3$,

$$\sqrt{\epsilon}\left(n_\epsilon(a,b) - \frac{\alpha_n m_n^-(a,b) + o(1)}{\epsilon} \right) \underset{\epsilon \to 0}{\Rightarrow} N(0, \alpha_n v_n^-(a,b)) \tag{6.14}$$

(2) If $a = 0$, $b > 0$ and $n = 2$,

$$\sqrt{\frac{\epsilon}{|\ln \epsilon|}}\left(n_\epsilon(0,b) - \frac{\alpha_n m_n^-(a,b) + o(1)}{\epsilon} \right) \underset{\epsilon \to 0}{\Rightarrow} N\left(0, \frac{\pi}{\beta^2}\right) \tag{6.15}$$

(3) If $a = 0$, $b > 0$ *and* $n = 1$,

$$\epsilon\left(n_\epsilon(0,b) - \frac{2}{\beta\epsilon}\ln(1-e^{-\beta b}) - \frac{2}{\beta\epsilon}|\ln(\beta\epsilon)|\right) \underset{\epsilon\to 0}{\Rightarrow} 2e^{-\frac{\beta x}{2}}K_1(2e^{-\frac{\beta x}{2}}). \tag{6.16}$$

Note that (6.16) is entirely free of the global parameters of Def. 6.5.

Proof: We first show the hypotheses of Theorem 6.2 to be satisfied in cases (1) and (2). Let N_ϵ and N'_ϵ correspond as before to A_ϵ and A'_ϵ , and

$$E_1(\epsilon) = \sqrt{\epsilon}|\ln\epsilon|^{\frac{1}{4}}, \qquad E_2(\epsilon) \equiv \infty.$$

In case (1) (if $a = 0$), the lowest eigenvalue of A'_ϵ is

$$E'_{\epsilon 1} = O(\epsilon^{\frac{1}{n}}),$$

while

$$N'_\epsilon(0, E_1(\epsilon)) = O\left(\frac{E_1{}^2(\epsilon)}{\epsilon}\right) = O\left(\sqrt{|\ln\epsilon|}\right). \tag{6.17}$$

Hence

$$\mathcal{V}(n'_\epsilon(0, E_1(\epsilon))) \le O\left(\epsilon^{-\frac{2}{n}}\sqrt{|\ln\epsilon|}\right). \tag{6.18}$$

In case (2),

$$\mathcal{V}(n'_\epsilon(0, E_1(\epsilon))) \le O\left(\epsilon^{-1}\sqrt{|\ln\epsilon|}\right); \tag{6.19}$$

Equations (6.18-19) prove (6.14). Lemma 6.4 gives (6.15) and (6.16); hence the first two cases follow from Theorem 6.2.

If $n = 1$ and $a = 0$, A_ϵ has spectrum

$$\mathcal{F}_\epsilon = \{\epsilon, \epsilon, 2\epsilon, 2\epsilon, 3\epsilon, 3\epsilon, \ldots\}. \tag{6.20}$$

Hence by Theorem 5.3 and the double multiplicities, the left side of (6.15) converges to the convolution of $e^{-e^{-\beta x}}$ with itself, which is on the right.∎

If $n_\epsilon = n_\epsilon(0,\infty)$, the distributions are explicitly

$$\sqrt{\epsilon}\left(n_\epsilon - \frac{n!\varsigma(n)\pi^{\frac{n}{2}} + o(1)}{\beta^n\Gamma(\frac{n+2}{2})\epsilon}\right) \underset{\epsilon\to 0}{\Rightarrow} N\left(0, \frac{n!\varsigma(n-1)\pi^{\frac{n}{2}}}{\beta^n\Gamma(\frac{n+2}{2})}\right) \tag{6.21}$$

f $n \ge 3$;

$$\sqrt{\frac{\epsilon}{|\ln\epsilon|}}\left(n_\epsilon - \frac{\pi^3 + o(1)}{3\beta^2\epsilon}\right) \underset{\epsilon\to 0}{\Rightarrow} N\left(0, \frac{\pi}{\beta^2}\right)$$

if $n = 2$; and

$$\epsilon\left(n_\epsilon - \frac{2}{\beta\epsilon}|\ln(\beta\epsilon)|\right) \underset{\epsilon\to 0}{\Rightarrow} 2e^{\frac{-\beta x}{2}} K_1\left(2e^{\frac{-\beta x}{2}}\right)$$

if $n = 1$.

For $0 \leq a < b < \infty$, let $\Delta_\epsilon(a,b)$ represent the total energy of bosons in the above ensemble whose energies lie between a and b.

THEOREM 6.7: *In a boson ensemble in $n+1$ dimensional Minkowski space, the asymptotic energy distribution is given by*

$$\sqrt{\epsilon}\left(\Delta_\epsilon(a,b) - \frac{\alpha_n m^-_{n+1}(a,b) + o(1)}{\epsilon}\right) \underset{\epsilon\to 0}{\Rightarrow} N(0, \alpha_n v^-_{n+2}(a,b)) \tag{6.22}$$

Proof: The hypotheses of Theorem 6.2 can be verified as above for $\Delta_\epsilon(a,b)$; the result then follows from Theorem 3.14 and (6.12) .∎

DEFINITIONS 6.8: If $\Delta_\epsilon(a,b)$ denotes a symmetric or antisymmetric energy r.v.,

$$\mathcal{D}_\epsilon(a,b) = \frac{\mathcal{E}(\Delta_\epsilon(a,b))}{\mathcal{E}(\Delta_\epsilon(0,\infty))} \tag{6.23}$$

is the *energy distribution function* corresponding to $\Delta_\epsilon(a,b)$. If the limit

$$\mathcal{D}(a,b) = \lim_{\epsilon\to 0} \frac{\mathcal{E}(\Delta_\epsilon(a,b))}{\mathcal{E}(\Delta_\epsilon(0,\infty))}, \tag{6.24}$$

exists, it is the *asymptotic energy distribution* of the net. If $\mathcal{D}(0,E)$ is absolutely continuous,

$$d(E) = \frac{d}{dE}\mathcal{D}(0,E)$$

is the *asymptotic energy density of the net.*

Theorem 6.7 gives the asymptotics of $\epsilon(\Delta_\epsilon(a,b))$: if $\Delta_\epsilon = \Delta_\epsilon(0,\infty)$, then

$$\sqrt{\epsilon}\left(\Delta_\epsilon - \frac{n n! \varsigma(n+1)\pi^{\frac{n}{2}} + o(1)}{\epsilon\beta^{n+1}\Gamma(\frac{n+2}{2})}\right) \underset{\epsilon\to 0}{\Rightarrow} N(0,v),$$

where

$$v = \frac{n(n+1)!\varsigma(n+1)\pi^{\frac{n}{2}}}{\beta^{n+2}\Gamma(\frac{n+2}{2})}. \tag{6.25}$$

COROLLARY 6.9: *The asymptotic energy density of the Minkowski $n+1$-space boson canonical ensemble is*

$$d_B(E) = \frac{E^n \beta^{n+1}}{(e^{\beta E} - 1) n! \varsigma(n+1)} \qquad (E \geq 0). \tag{6.26}$$

We now find the asymptotic fermion distributions corresponding to A_ϵ.

THEOREM 6.10: *Let $n_\epsilon(a,b)$ be the fermion number in the Minkowski space canonical ensemble. Then*

$$\sqrt{\epsilon}\left(n_\epsilon(a,b) - \frac{\alpha_n m_n^+(a,b) + o(1)}{\epsilon}\right) \underset{\epsilon \to 0}{\Rightarrow} N(0, \alpha_n v_n^+(a,b)). \tag{6.27}$$

Similarly, for fermion energy,

$$\sqrt{\epsilon}\left(\Delta_\epsilon(a,b) - \frac{\alpha_n m_{n+1}^+(a,b) + o(1)}{\epsilon}\right) \underset{\epsilon \to 0}{\Rightarrow} N(0, \alpha_n v_{n+2}^+(a,b)). \tag{6.28}$$

Hence if $\Delta_\epsilon = \Delta_\epsilon(0,\infty)$

$$\sqrt{\epsilon}\left(\Delta_\epsilon - \frac{n! n \varsigma(n+1) \pi^{\frac{n}{2}} (1 - 2^{-n}) + o(1)}{\epsilon \beta^{n+1} \Gamma(\frac{n+2}{2})}\right) \underset{\epsilon \to 0}{\Rightarrow} N(0, v), \tag{6.29}$$

with

$$v = \frac{n(n+1)! \varsigma(n+1) \pi^{\frac{n}{2}} (1 - 2^{-n})}{\beta^{n+2} \Gamma(\frac{n+2}{2})}. \tag{6.30}$$

COROLLARY 6.11: *The fermion ensemble in Minkowski $n+1$-space has asymptotic energy density*

$$d_F(E) = \frac{E^n \beta^{n+1}}{(e^{\beta E} + 1) n! \varsigma(n+1)(1 - 2^{-n})}. \tag{6.31}$$

§6.4 Distributions in Einstein Space

We now consider distributions in two and four dimensional Einstein space, $U^2 = S^1 \times \mathbf{R}$ and $U^4 = S^3 \times \mathbf{R}$. In four dimensions the single particle Hamiltonian A_ϵ is the square root of a constant perturbation of the Laplace-Beltrami operator for which the wave equation satisfies Huygens' principle.

We first consider Einstein 2-space, with

$$\epsilon = \frac{1}{R}, \tag{6.32}$$

where R is the radius of the spatial portion S^1. Clearly A_ϵ has spectrum (6.20), and the $R \to \infty$ asymptotics follow directly from Theorems 6.6-6.10.

Since the discrete pre-asymptotic operator A_ϵ is of more direct interest here, we will analyze associated distributions more carefully. We define two combinatorial functions.

DEFINITIONS 6.12: If $0 \leq a < b \leq \infty$, and $l \in \mathbf{Z}^+$, then

$$p_l^{(2)}(a,b) \equiv \sum_{n(\cdot)\in P_l(a,b)} \prod_{a\leq i\leq b} (n(i)+1) \tag{6.33}$$

where

$$P_l(a,b) = \left\{ n(\cdot) \in \mathcal{N}_{[a,b]} : \sum_{a\leq i\leq b} in(i) = l \right\}, \tag{6.34}$$

and $\mathcal{N}_{[a,b]}$ is the set of all functions $n(\cdot) : \mathbf{Z}^+ \cap [a,b] \to \mathbf{Z}^+$. Define $q_l^{(2)}(a,b)$ as the number of ways of expressing $l \in \mathbf{Z}^+$ as a sum of integers in the interval $[a,b]$, no single integer being used more than *twice*.

We present the following without proof.

THEOREM 6.13: *If $0 \leq a < b \leq \infty$ and $\Delta_\epsilon(a,b)$ is the boson energy in Einstein 2-space at inverse temperature $\beta > 0$, then $\Delta_\epsilon(a,b)$ is concentrated on*

$$\epsilon\mathbf{Z}^+ = \{\epsilon j : j \in \mathbf{Z}^+\}, \tag{6.35}$$

and

$$P(\Delta_\epsilon(a,b) = \epsilon l) = \frac{p_l^{(2)}(a,b)e^{-\epsilon\beta l}}{K_B} \qquad (l \in \mathbf{Z}^+), \tag{6.36}$$

where

$$K_B = \prod_{j=1}^{\infty} (1 - e^{-\beta\epsilon j})^{-2}. \tag{6.37}$$

If Δ_ϵ represents fermion energy, then $p_l^{(2)}$ is replaced by $q_l^{(2)}$, and K_B by

$$K_F = \prod_{j=1}^{\infty} (1 + e^{-\beta\epsilon j})^2. \tag{6.38}$$

The $R \to \infty$ limits of the two-dimensional distributions are specializations of the $n = 1$ cases in §6.1.

PROPOSITION 6.14: *As $\frac{1}{\epsilon} = R \to \infty$, the asymptotic energy densities of bosons and fermions in Einstein 2-space are*

$$d_{B_2}(E) = \frac{6E\beta^2}{(e^{\beta E} - 1)\pi^2}; \qquad d_{F_2}(E) = \frac{12E\beta^2}{(e^{\beta E} + 1)\pi^2}.$$

We now consider distributions in Einstein 4-space $S^3 \times \mathbf{R}$. The Hamiltonian A_ϵ acts as $\frac{1}{i}\frac{\partial}{\partial \tau}$ (where τ is time) on the Hilbert space of solutions of the invariant non-self-interacting scalar wave equation. It has spectrum $\sigma(A_\epsilon) = \{n\epsilon : n \in \mathrm{N}\}$, $n\epsilon$ having multiplicity n^2.

PROPOSITION 6.15: *If $\beta > 0$ then $e^{-\beta d\Gamma_B(A_\epsilon)}$ and $e^{-\beta d\Gamma_F(A_\epsilon)}$ are trace class.*

Proof: By Proposition 1.4 it suffices to show $e^{-\beta A_\epsilon}$ is trace class; we have

$$\mathrm{tr}\, e^{-\beta A_\epsilon} = \sum_{j=1}^{\infty} j^2 e^{-\beta\epsilon j} = e^{\beta\epsilon}\frac{e^{\beta\epsilon}+1}{(e^{\beta\epsilon}-1)^3} < \infty. \blacksquare \tag{6.39}$$

Note that

$$\begin{aligned} K_B &= \mathrm{tr}\, e^{-\beta d\Gamma_B(A_\epsilon)} = \prod_{j=1}^{\infty}(1-e^{-\beta\epsilon j})^{-j^2} \\ K_F &= \mathrm{tr}\, e^{-\beta d\Gamma_F(A_\epsilon)} = \prod_{j=1}^{\infty}(1+e^{\beta\epsilon j})^{j^2} \end{aligned} \tag{6.40}$$

For completeness we derive the exact distribution of energy in terms of the following integer-valued combinatorial functions.

DEFINITIONS 6.16: If $0 \le a < b \le \infty$ and $l \in \mathbf{Z}^+$, let

$$p_l^{(j^2)}(a,b) = \sum_{n(\cdot)\in R_l(a,b)} \left(\prod_{a\le j\le b} \binom{n(j)+j^2-1}{j^2-1} \right), \tag{6.41}$$

where

$$R_l(a,b) = \left\{ n(\cdot) \in \mathcal{N}_{[a,b]} : \sum_{a\le i\le b} n(i) = l \right\},$$

with $\mathcal{N}_{[a,b]}$ as in Definitions 6.12, and $\binom{c}{d}$ denoting c choose d. Let $q_l^{(j^2)}(a,b)$ be the number of ways (without regard to order) of expressing $l \in \mathbf{Z}^+$ as a sum of integers in $[a,b]$, each integer j being used no more than j^2 times.

THEOREM 6.17: *The Bose energy $\Delta_\epsilon(a,b)$ in Einstein 4-space at inverse temperature $\beta > 0$, is concentrated on $\epsilon\mathbf{Z}^+$, and*

$$\mathcal{P}(\Delta_\epsilon(a,b) = \epsilon l) = \frac{p_l^{(j^2)}(a,b)e^{-\beta\epsilon l}}{K_B} \qquad (l \in \mathbf{Z}^+); \tag{6.42}$$

in the Fermi case $p_l^{(j^2)}$ is replaced by $q_l^{(j^2)}$, and K_B by K_F.

We now consider asymptotics of Einstein space distributions. Denote by A'_ϵ the ϵ-discrete operator corresponding to the spectral measure μ for which $F(E) = \frac{E^3}{6}$. Let $N_\epsilon(0,E)$ be the cardinality of $\sigma(A_\epsilon) \cap (0,E)$, with the analogous definition for $N'_\epsilon(0,E)$ relative to A'_ϵ.

LEMMA 6.18: *If $\epsilon > 0$ and $E \geq 0$, then $N_\epsilon(0,E) \leq N'_{\epsilon^3}(0,2E)$.*

Proof: We have

$$N'_{\epsilon^3}(0,2E) = \left[\frac{4E^3}{3\epsilon^2}\right],$$

while

$$N_\epsilon(0,E) = \sum_{j=1}^{\left[\frac{E}{\epsilon}\right]} j^2 = \frac{1}{6}\left[\frac{E}{\epsilon}\right]\left(\left[\frac{E}{\epsilon}\right]+1\right)\left(2\left[\frac{E}{\epsilon}\right]+1\right). \tag{6.43}$$

We have

$$\left[\frac{E}{\epsilon}\right] \leq \left[\frac{E}{\epsilon}\right]^2 \leq \left[\frac{E}{\epsilon}\right]^3 \leq \left[\frac{4E^3}{3\epsilon^3}\right] \qquad (\epsilon > 0,\ E \geq 0), \tag{6.44}$$

and equations (6.42-43) imply the result.∎

LEMMA 6.19: *Let $n_\epsilon(0,E)$ be the boson number corresponding to A_ϵ in Einstein 4-space. Then*

$$\epsilon^3 \mathcal{V}(n_\epsilon(0,\sqrt{\epsilon})) \underset{\epsilon\to 0}{\to} 0 \tag{6.45}$$

and

$$\sup_{E \geq \sqrt{\epsilon}} \left| \frac{N_\epsilon(0,E)}{N'_{\epsilon^3}(0,E)} - 1 \right| \underset{\epsilon\to 0}{\to} 0. \tag{6.46}$$

Proof: We have

$$\begin{aligned} \mathcal{V}(n'_{\epsilon^3}(0,\sqrt{\epsilon})) &= \sum_{j=1}^{\left[\frac{\sqrt{\epsilon}^3}{6\epsilon^3}\right]} \frac{e^{\beta\epsilon(6j)^{\frac{1}{3}}}}{(e^{\beta\epsilon(6j)^{\frac{1}{3}}}-1)^2} \leq e^{\beta\sqrt{\epsilon}} \sum_{j=1}^{\left[\frac{1}{6\epsilon\sqrt{\epsilon}}\right]} \frac{1}{(\beta\epsilon(6j)^{\frac{1}{3}})^2} \\ &\leq \frac{e^{\beta\sqrt{\epsilon}}}{(\beta\epsilon)^2} \int_0^{\frac{1}{\epsilon\sqrt{\epsilon}}} x^{-\frac{2}{3}}\,dx \\ &= \frac{3e^{\beta\epsilon}\epsilon^{-\frac{5}{2}}}{\beta^2}, \end{aligned} \tag{6.47}$$

from which (6.45) follows. Equation (6.46) is implied by

$$N'_{\epsilon^3}(0,E) = \left[\frac{E^3}{6\epsilon^3}\right] \tag{6.48}$$

and (6.43).

LEMMA 6.20: *If* $0 \leq a < b \leq \infty$, *then*

$$\frac{N_\epsilon(a,b)}{N'_{\epsilon^3}(a,b)} \underset{\epsilon\to 0}{\to} 1,$$

and

$$\epsilon^3 \mathcal{V}(n'_\epsilon(E,\infty)) \underset{\epsilon\to\infty}{\to} 0 \tag{6.49}$$

uniformly in ϵ.

Proof: The first assertion follows from (6.46). The second follows from the existence of fixed $\epsilon^*, E^* > 0$ such that

$$N_\epsilon(E,E') \leq 2N'_{\epsilon^3}\left(\frac{E}{2}, E_1\right) \qquad (\epsilon \leq \epsilon^*,\ E^* \leq E < E_1 < \infty),$$

and the fact that

$$\epsilon^3 \mathcal{V}(n'_{\epsilon^3}(E,\infty)) \underset{E\to\infty}{\to} 0 \tag{6.50}$$

uniformly in ϵ.∎

THEOREM 6.21: *The boson number in Einstein 4-space satisfies*

$$\sqrt{\epsilon^3}\left(n_\epsilon(a,b) - \frac{m(\epsilon)}{\epsilon^3}\right) \underset{\epsilon\to 0}{\to} N(0,v),$$

where

$$m(\epsilon) = \frac{1}{2}\int_a^b \frac{E^2}{e^{\beta E}-1}dE + o(1) \qquad (\epsilon \to 0) \tag{6.51}$$

and

$$v = \frac{1}{2}\int_a^b \frac{E^2 e^{\beta E}}{(e^{\beta E}-1)^2}dE.$$

If n_ϵ *represents fermion number, each "–" is replaced by "+".*

Proof: This follows in the Bose case from Lemmas 6.18-20, Theorem 6.1 and its Corollary, and Theorem 3.14.∎

THEOREM 6.22: *The Einstein 4-space asymptotic boson and fermion number and energy densities are identical to those in Minkowski 4-space. Precisely, equations (6.14), (6.22), and (6.27-28) still hold.*

COROLLARY 6.23: *The asymptotic energy densities of Bose and Fermi ensembles in Einstein 4-space are*

$$d_{B_4}(E) = \frac{15E^3\beta^4}{(e^{\beta E}-1)\pi^4}$$

$$d_{F_4}(E) = \frac{120E^3\beta^4}{7(e^{\beta E}+1)\pi^4}.$$

§ 6.5 Physical Discussion

The requirements of localization in physical (as well as momentum) space result in the approximation in §6.2 of the Hamiltonian for massless scalar particles in Minkowski space. There seems to be no direct way of discretizing $d\Gamma(A)$ without reference to A. The r.v. n_ϵ is interpreted as boson number in an n-dimensional torus of volume $V = \frac{(2\pi)^n}{\epsilon}$. If $n \geq 2$ the expected number per unit volume is asymptotically $\frac{\alpha_n m_n^-}{(2\pi)^n}$; the same density occurs for Einstein 4-space (Theorem 6.21). If $n = 1$ the mean density of photons in the energy interval $[0, b]$ is asymptotically $\frac{|\ln(\beta\epsilon)|}{\pi\beta}$ $(\epsilon \to 0)$, and thus large (and R-dependent) in Einstein space and infinite in Minkowski space. Since asymptotic density is independent of b, the divergence clearly arises from bosons with effectively vanishing energy; this is an effect of Bose condensation, in which large numbers of particles appear in the lowest energy levels. The corresponding spatial energy density is, however, finite according to Theorem 6.7, which gives mean density

$$D_n = \frac{n}{2^n\pi^{\frac{n}{2}}\Gamma(\frac{n+2}{2})}\int_0^{E_2}\frac{E}{e^{\beta E}-1}dE.$$

This divergent density of low-energy bosons with finite corresponding energy density has an analog in the "infrared catastrophe" of quantum electrodynamics (see [BD], §17.10), in which an infinite number of "soft photons" with finite total energy is emitted by an electrical current. The extreme density is correlated with the lack of normality in photon numbers.

The energy distribution of bosons in a Minkowski $n+1$-space canonical ensemble in (6.25) is the "Planck law" for such a system. An observer of scalar photons measures the proportion of photon energy in the frequency interval $[\nu, \nu + \Delta\nu]$ to be $\int_{h\nu}^{h(\nu+\Delta\nu)} d_B(E)dE$, with h Planck's constant. Analogously, (6.31) gives the corresponding law for fermions; in an ensemble consisting of neutrinos, the proportion of total neutrino energy observed in $[E, E+\Delta E]$ is thus $\int_E^{E+\Delta E} d_F(E)dE$. Note that the Planck laws for Einstein 4-space coincide asymptotically with those in Minkowski space; this obviously also holds in one dimension. This indicates that the cosmic background radiation expected in an approximately steady-state model of the universe is largely independent of the underlying manifold. The specific correspondences in this chapter are a consequence of the physical identity of Minkowski space and the "$R \to \infty$ limit" of Einstein space of radius R.

CHAPTER 7

THE LEBESGUE INTEGRAL

In this chapter we introduce a natural and useful generalization of the notions in Chapter 3. Lebesgue integration of r.v.-valued functions on a measure space is the maximal completion of Riemann integration. The step from Riemann to Lebesgue integration shifts the focus from the domain to the range of the integrated function; indeed, the ordinary Lebesgue integral is a Riemann integral of the identity function on the range space with respect to the domain-induced measure; this viewpoint will be used here.

The present integration theory is in fact interpretable as a formal extension of the theory of semi-stable stochastic processes (see [La, BDK]), with an abstract measure space replacing time. The r.v.-valued function X being integrated yields an r.v.-valued measure $\mathcal{M}$ defined by $\mathcal{M}(A) =$ integral of X over A; this measure is clearly a generalized stochastic process. Indeed, in the notation of this chapter, if $\{X(t)\}_{t \in \mathbf{R}}$ is an collection of independent standard normal r.v.'s, then $\int_0^t X(t)(dt)^{1/2}$ is simply Brownian motion.

The advantage of the present approach to that of standard integrals of distribution-valued random functions (e.g., "white noise") [Che,R1,V] is that it does not require the existence of a metric (or smooth volume element) on the underlying measure space. Precisely, the present integration theory would, on a Riemmanian manifold, be equivalent to (linear) integration of an r.v.-valued distribution with covariance

$$\mathcal{E}(X(\lambda_1)X(\lambda_2)) = \mathcal{V}(X(\lambda_1))\delta(\lambda_1, \lambda_2) \qquad (\lambda_i \in \Lambda)$$

(Here, δ denotes the point mass at λ_2, in the variable λ_1.) However, an abstract measure space generally has no such object.

A novel aspect of the Lebesgue integral is the use of a non-linear volume element $\phi(d\mu)$. This may seem rather ad hoc; but in Chapter 8 the Lebesgue integral is shown to be isomorphic to a linear integral over functions with range in a space of logarithms of characteristic functions.

Lebesgue integration is the natural environment for detailed study of integrals of independent random variables; however, aside from proof of associated fundamentals which will comprise most of this and the next chapter, the approach here will be relatively goal and applications oriented. Further on, the measure space $(\Lambda, \mathcal{B}, \mu)$ will be the spectrum of an abelian von Neumann algebra of quantum observables.

The material in the next three chapters will be largely independent of previous material, and the probabilistic content stands on its own. The proofs may be omitted on a first reading.

§ 7.1 Definitions and Probabilistic Background

We will often deal with normalized integrals (sums) of random variables with infinite variance; the resulting limits will depend strongly on the tails of the integrated distributions. We recall the bare essentials of general limit theorems for sums of independent random variables; see [GK] for details.

In order to formulate the most general central limit results, we define the Lévy-Khinchin transform of an infinitely divisible distribution. Recall every such distribution ν has characteristic function $\Phi = e^{\psi}$, where ψ is continuous and vanishes at $t = 0$. The book of Loèwe [Lo] has a proof of

THEOREM 7.1: *The distribution ν is infinitely divisible if and only if $\Phi = e^{\psi}$, where ψ is given (uniquely) by*

$$\psi(t) = i\gamma t + \int_{-\infty}^{\infty}\left(e^{itx} - 1 - \frac{itx}{1+x^2}\right)\frac{1+x^2}{x^2}dG(x), \tag{7.1}$$

with $\gamma \in \mathbf{R}$ and G is a non-negative multiple of a d.f.

The pair (γ, G) is the *Lévy-Khinchin transform* of ν. Note this is additive, in that if ν_i has transform (γ_i, G_i) $(i = 1, 2)$, then $\nu_1 * \nu_2$ has transform $(\gamma_1 + \gamma_2, G_1 + G_2)$. It can be shown that the transform is continuous from the space of distributions ν in the topology of weak convergence, to the pairs (γ, G) in the topology of $\mathbf{R}^1$ crossed with the topology of weak convergence.

Let $\{X_{nk}\}$, $1 \le n < \infty$, $1 \le k \le k_n$ be a double array of independent random variables; $k_n = \infty$ is allowed.

DEFINITION 7.2: The variables X_{nk} are *infinitesimal* if for every $\epsilon > 0$,

$$\sup_{1 \le k \le k_n} P(|X_{nk}| \ge \epsilon) \underset{n\to\infty}{\to} 0.$$

For any monotone functions $G_n(x)$, $G(x)$ of bounded variation, we write $G_n(x) \Rightarrow G(x)$ if the same is true of the corresponding measures on $\mathbf{R}^1$.

LEMMA 7.3.1: *Let $\{X_n\}$ be a sequence of independent r.v.'s. In order that $\sum_{n=1}^{\infty} X_n$ converge weakly and order-independently, it is necessary and sufficient that*

$$\sum_{k=1}^{\infty} \mathcal{E}\left(\frac{X_n^2}{1+X_n^2}\right), \qquad \sum_{k=1}^{\infty} \mathcal{E}\left(\frac{X_n}{1+X_n^2}\right) \tag{7.2}$$

converge absolutely.

Proof: By the three series theorem, order-independent convergence above is for any $\tau > 0$ equivalent to absolute convergence of the series

$$\sum_n \int_{|x|\geq\tau} dF_n(x), \quad \sum_n \int_{|x|\leq\tau} x\,dF_n(x), \quad \sum_n \int_{|x|\leq\tau} x^2 dF_n(x), \tag{7.3}$$

where F_n is the d.f. of X_n. But absolute convergence of (7.3) implies convergence of the first series in (7.2). Thus proving absolute convergence of the second series reduces to the same for

$$\sum_n \int_{|x|\leq\tau} \left(\frac{x}{1+x^2} - x\right) dF_n(x) = -\sum_n \int_{|x|\leq\tau} \frac{x^3}{1+x^2} dF_n(x) \tag{7.4}$$

Convergence of the latter follows from that of the first series in (7.2).

Conversely, if (7.2) converge absolutely, so do the first and third series of (7.3), while convergence of the middle series follows from a subtraction argument, as in (7.4).∎

Henceforth, $F_{nk}(x)$ will denote the (cumulative) distribution function of X_{nk}, and all infinite sums must converge order-independently to be well defined. Assume that $\{X_{nk}\}_{k=1}^{kn}$ form an infinitesimal array. The proof of the following theorem for k_n finite is in [Lo]; we extend it to general k_n.

THEOREM 7.3: *Let $S_n = \sum_k X_{nk}$. In order that S_n converge weakly to a distribution S, it is necessary and sufficient that*

$$G_n \Rightarrow G, \qquad \gamma_n \to \gamma. \tag{7.5}$$

Here,

$$\gamma_n = \sum_{k=1}^{k_n} a_{nk} + \int_{-\infty}^{\infty} \frac{y}{1+y^2} d\overline{F}_{nk}(y), \qquad G_n(x) = \sum_{k=1}^{k_n} \int_{-\infty}^{x} \frac{y^2}{1+y^2} d\overline{F}_{nk}(y), \tag{7.6}$$

and

$$a_{nk} = \int_{|y|\leq\tau} y\, dF_{nk}(y), \qquad \overline{F}_{nk}(x) = F_{nk}(x + a_{nk}),$$

with $\tau > 0$ any fixed constant. The pair (γ, G) is the Lévy-Khinchin transform of S.

Proof: This follows using approximation by finite sums. For example, if S_n converge weakly to S, the same is true of $S_n^* = \sum_{k=1}^{k_n^*} X_{nk}$, where k_n^* is finite but sufficiently large; using the result for S_n^* and letting k_n^* become infinite proves (7.5). The only difficulty lies in proving the sum defining S_n converges (order-independently) if and only if (7.6) does.

To this end, note that if S_n exists, the three series theorem implies $\sum_{k=1}^{\infty}|a_{nk}|$ converges. Letting $\chi(y) = \frac{y}{1+y^2}$,

$$\sum_{k=1}^{\infty}\int_{-\infty}^{\infty}\chi(y)\;d\bar{F}_{nk}(y) = \sum_{k=1}^{\infty}\int_{-\infty}^{\infty}\chi(y-a_{nk})dF_{nk}(y)$$
$$= \sum_{k=1}^{\infty}\int_{-\infty}^{\infty}(\chi(y) - a_{nk}\chi'(y_{nk}))dF_{nk}(y) \tag{7.7}$$

where $y - a_{nk} \leq y_{nk} \leq y$ is determined according to the mean value theorem. By infinitesimality of the X_{nk}, this sum converges if and only if $\sum_{k=1}^{\infty}\int_{-\infty}^{\infty}\chi(y)dF_{nk}(y)$ converges. This (again three series) occurs if and only if $\sum_{k=1}^{\infty}\int_{-\tau}^{\tau}\chi(y)dF_{nk}(y)$ converges for some $\tau > 0$. The latter follows from the Lemma. The existence of the limit $G(x)$ follows similarly.

Conversely, assume the limits (7.6) exist. Then the sum

$$\sum_{k=1}^{\infty} a_{nk} + \int_{-\tau}^{\tau} y d\bar{F}_{nk}(y)$$

converges for any $\tau > 0$, since $y - \chi(y)$ is dominated by $\frac{y^2}{1+y^2}$ for $|y| \leq \tau$. Therefore we have convergence of

$$\sum_{k=1}^{\infty} a_{nk} + \int_{-\tau+a_{nk}}^{\tau+a_{nk}}(y-a_{nk})\,dF_{nk}(y) = \sum_{k=1}^{\infty} a_{nk} + \sum_{k=1}^{\infty} a_{nk}P(|X_{nk}-a_{nk}| > \tau)$$
$$+ \sum_{k=1}^{\infty}\left(\int_{\tau+}^{\tau+a_{nk}} - \int_{-\tau}^{-\tau+a_{nk}^{-}}\right) y\,dF_{nk}(y) \tag{7.8}$$

Using infinitesimality of a_{nk} and finiteness of $\sum_k P(|X_{nk}| \geq C)$ for $C > 0$ (the latter follows from convergence of the second sum in (7.6)), the last two sums converge. Therefore $\sum_k a_{nk}$ is convergent, as is $\sum\int_{-\infty}^{\infty}\chi\,d\bar{F}_{nk}$. Using (7.7) and the infinitesimality of $\{a_{nk}\}$, we conclude that $\sum_k\int_{-\infty}^{\infty}\chi\,dF_{nk}$ converges (absolutely) as well. Similarly, $\sum_k\int_{-\infty}^{\infty}\frac{y^2}{1+y^2}dF_{nk} < \infty$. Thus, by Lemma 7.2.1, S_n exists and is order-independent.∎

Henceforth let

$$\chi^*(x) = \begin{cases} x; & |x| \leq 1 \\ \frac{1}{x}; & |x| \geq 1 \end{cases}, \qquad \chi(x) = \frac{x}{1+x^2}, \qquad \theta(\chi) = \frac{x^2}{1+x^2}. \tag{7.9}$$

COROLLARY 7.3.1: *In order that S_n converge weakly, it is necessary and sufficient that*

$$\gamma' = \lim_{n\to\infty}\sum_{k=1}^{k_n} b_{nk}, \quad G(x) = \text{w-}\lim\sum_{k=1}^{k_n}\int_{-\infty}^{x}\theta(y)d\tilde{F}_{nk}(y)$$

exist, where

$$b_{nk} = \int_{-\infty}^{\infty} \chi^*(y) dF_{nk}(y), \qquad \tilde{F}_{nk}(y) = F_{nk}(y + b_{nk}) \tag{7.10}$$

The weak limit S has Lévy-Khinchin transform $(\gamma, G(u))$, where

$$\gamma = \lim_{n\to\infty} \sum_{k=1}^{k_n} b_{nk} + \int_{-\infty}^{\infty} \chi(y) d\tilde{F}_{nk}(y).$$

Proof: If S_n converges weakly, then $G_n(x) = \sum_k \int_{-\infty}^{x} \theta(y) d\overline{F}_{nk}(y) \underset{n\to\infty}{\Rightarrow} G^*(x)$ for some multiple G^* of a d.f. If $\pm\tau$ are continuity points of G^*, then

$$\sum_k \int_{-\infty}^{-\tau} d\overline{F}_{nk}(y) = \int_{-\infty}^{-\tau} \frac{1+y^2}{y^2} dG_n(y) \tag{7.11}$$

converges as $n \to \infty$. Similarly, $\sum_k \int_{|y|\geq\tau} d\overline{F}_{nk}(y)$ converges; and

$$\sum_k \int_{|y|\leq\tau} \frac{y^3}{1+y^2} d\overline{F}_{nk}(y) = \int_{|y|\leq\tau} y dG_n(y)$$

converges as well, as does $\sum_k \int_{|y|\geq\tau} \frac{y}{1+y^2} d\overline{F}_{nk}(y)$. Thus, by convergence of γ_n in (7.6) and the identity $x - \frac{x}{1+x^2} = \frac{x^3}{1+x^2}$, we have convergence of

$$\sum_{k=1}^{k_n} \left(a_{nk} + \int_{|y|\leq\tau} y d\overline{F}_{nk}(y) \right),$$

with a_{nk} as in the Theorem. We conclude the convergence of

$$\begin{aligned} \sum_{k=1}^{k_n} a_{nk} + \int_{-\tau-a_{nk}}^{\tau-a_{nk}} y \, d\overline{F}_{nk}(y) &= \sum_{k=1}^{k_n} \left(a_{nk} + \int_{|y|\leq\tau} (y - a_{nk}) \, dF_{nk}(y) \right) \\ &= \sum_{k=1}^{k_n} \left(a_{nk} + a_{nk} \int_{|y|\geq\tau} dF_{nk} \right); \end{aligned} \tag{7.12}$$

the sum $\sum_{k=1}^{k_n} a_{nk} \int_{|y|\geq\tau} dF_{nk}$ clearly converges to 0 as $n \to \infty$ by infinitesimality of the a_{nk}, and hence $\sum_{k=1}^{k_n} a_{nk}$ converges to a limit as $n \to \infty$.

On the other hand, since for $x < 0$, $\sum \overline{F}_{nk}(x)$ converges weakly (eq. (7.11)),

$$\int_{-\infty}^{-\tau-a_{nk}} \chi^*(x + a_{nk}) \, d\left(\sum_k \overline{F}_{nk}(x) \right) = \sum_k \int_{-\infty}^{-\tau} \chi^*(x) dF_{nk}$$

converges, so that the sum defining γ' converges (the sum $\sum_k \int_\tau^\infty \chi^* dF_{nk}$ converges by similar arguments). Let $c_{nk} \equiv b_{nk} - a_{nk}$, and consider

$$\begin{aligned}\sum_k \int_{-\infty}^{x} \theta(y)\, dF_{nk}(y + b_{nk}) &= \sum_k \int_{-\infty}^{x-c_{nk}} \theta(y + c_{nk})\, d\overline{F}_{nk}(y) \\ &= \sum_k \left(\int_{-\infty}^{x-c_{nk}} \theta(y)\, d\overline{F}_{nk} + \int_{-\infty}^{u-c_{nk}} c_{nk}^* \theta'(y)\, d\overline{F}_{nk} \right),\end{aligned} \tag{7.13}$$

where c_{nk}^* is between 0 and c_{nk}. Without loss of generality, we let $\tau < 1$; then

$$\sum_k |c_{nk}| = \sum_k |b_{nk} - a_{nk}| = \sum_k \left| \int_{|x| \geq \tau} \chi^*(y)\, dF_{nk}(y) \right| \tag{7.14}$$

remains bounded as $n \to \infty$, as does $\sum |c_{nk}^*|$. We have $\int_{-\infty}^{\infty} |\theta'(y)| d\overline{F}_{nk} \underset{n \to \infty}{\to} 0$ uniformly in k, so that the last term on the right of (7.13) vanishes as $n \to \infty$, and

$$\sum_k \int_{-\infty}^{x} \theta(y) d\tilde{F}_{nk}(y) \Rightarrow G^*(u) = G(u). \tag{7.15}$$

Conversely, if the series for γ' and $G(u)$ converge absolutely in k and as $n \to \infty$, then it follows along similar lines that $\sum_{k=1}^{k_n} \int_{|y| \geq \tau} \chi^*(y) dF_{nk}(y)$ converges as $n \to \infty$, for any $\tau > 0$. Thus, since $\sum_{k=1}^{k_n} \chi^* dF_{nk}$ converges as $n \to \infty$, $\Sigma_k a_{nk}$ converges as well. Using by now standard arguments, it follows that $\sum_k \int_{-\infty}^{\infty} \chi(y) d\overline{F}_{nk}(y)$ converges as $n \to \infty$, as does $\sum \left(a_{nk} + \int_{-\infty}^{\infty} \chi(y) d\overline{F}_{nk}(y)\right)$. Similarly, if x is a continuity point of $G(x)$, then $\sum_k \int_{-\infty}^{x} \theta(y) d\overline{F}_{nk}(y) \to G(x)$.

We now prove the last statement of the Corollary. We have

$$\begin{aligned}&\lim_{n \to \infty} \sum_k \left[a_{nk} + \int_{-\infty}^{\infty} \chi\, d\overline{F}_{nk} \right] = \\ &\lim_{n \to \infty} \sum_k \left[a_{nk} + \int_{-\infty}^{\infty} \left(\frac{\chi - \chi^*}{\theta} \right) \theta d\tilde{F}_{nk} + \int_{-\infty}^{\infty} \chi^* d\overline{F}_{nk} \right],\end{aligned} \tag{7.16}$$

since the measures $\sum_k \theta d\overline{F}_{nk}$ and $\sum_k \theta d\tilde{F}_{nk}$ have the same weak limit (see (7.15)). By the mean value theorem,

$$\begin{aligned}\sum_k \int_{-\infty}^{\infty} \chi^* d\tilde{F}_{nk} &= \sum_k \int_{-\infty}^{\infty} \chi^* dF_{nk} - b_{nk} \int_{-\infty}^{\infty} \chi^{*\prime}(y - b_{nk}^*(y)) dF_{nk} \\ &= \sum_k b_{nk} \left(\int_{-\infty}^{\infty} (1 - \chi^{*\prime}(y - b_{nk}^*(y)) dF_{nk}) \right).\end{aligned} \tag{7.17}$$

Since $\sum_k \int_{-\infty}^{\infty} (1-\chi^{*\prime}(y-b_{nk}^*))dF_{nk}$ is uniformly bounded (since $\sum_k \int_{|y|\geq 1} dF_{nk}$ is), and $b_{nk} \underset{n\to\infty}{\to} 0$ uniformly, (7.17) vanishes as $n \to \infty$. Similarly, if τ is a continuity point of G and

$$\chi^{**}(x) = \begin{cases} x; & |x| \leq \tau \\ 0; & |x| > \tau, \end{cases}$$

then $\displaystyle\sum_k \int_{-\infty}^{\infty} \chi^{**} d\overline{F}_{nk} \underset{n\to\infty}{\to} 0$ uniformly. Thus (7.16) is given by

$$\begin{aligned} &\lim_{n\to\infty} \sum_k \left[a_{nk} + \int_{-\infty}^{\infty} (\chi^* - \chi^{**})\, d\overline{F}_{nk} + \int_{-\infty}^{\infty} \chi\, d\tilde{F}_{nk}\right] \\ &= \lim_{n\to\infty} \sum_k \left[a_{nk} + \int_{-\infty}^{\infty} (\chi^* - \chi^{**})\, d\tilde{F}_{nk} + \int_{-\infty}^{\infty} \chi\, d\tilde{F}_{nk}\right] \\ &= \lim_{n\to\infty} \sum_k \left[a_{nk} + \int_{|y|>\tau} \frac{1}{y}\, d\tilde{F}_{nk} + \int_{-\infty}^{\infty} \chi\, d\tilde{F}_{nk}\right] \qquad (7.18) \\ &= \lim_{n\to\infty} \sum_k b_{nk} + \int_{-\infty}^{\infty} \chi\, d\tilde{F}_{nk}, \end{aligned}$$

the last equality following from

$$\sum_k \int_{|y|>\tau} \frac{1}{y}\, d\tilde{F}_{nk} - \int_{|y+b_{nk}|>\tau} \frac{1}{y}\, d\tilde{F}_{nk} \underset{n\to\infty}{\to} 0.$$

Equation (7.18), together with (7.15), completes the proof.∎

§ 7.2 Definition of the Lebesgue Integral

Since we will study integrals of measure-valued functions, we consider metrics on spaces of probability distributions. Let $\mathcal{D}$ be the set of probability Borel measures on $\mathbf{R}^1$, and $\mathcal{D}^*$ the set of finite Borel measures. Define the Lévy metric ρ^L by

$$\rho^L(\nu_1, \nu_2) = \inf\{h : F_1(x-h) - h \leq F_2(x) \leq F_1(x+h) + h\}, \qquad (\nu_1, \nu_2 \in \mathcal{D}^*), \qquad (7.19)$$

where F_i is the d.f. of ν_i. The Lévy metric is compatible with the topology of convergence in distribution on $\mathcal{D}$ (see, e.g., [GK]). Note that ρ^L can be defined for any pair ν_1, ν_2 of Borel measures on $\mathbf{R}$. This general definition will be used here.

We will also require a stronger metric which emphasizes tail properties of measures. Let ϕ: $\mathbf{R}^+ \to \mathbf{R}^+$ be defined for small arguments, and nonvanishing. Define (recall (7.9))

$$G_{\nu,\delta}(x) = \frac{1}{\delta} \int_{-\infty}^{x} \theta(y)\; d\nu\!\left(\frac{y}{\phi(\delta)}\right). \qquad (7.20)$$

We define

$$\rho_{\phi,\delta}(\nu_1,\nu_2) = \rho^L(G_{\nu_1,\delta}, G_{\nu_2,\delta}) + \frac{1}{\delta}\left|\int_{-\infty}^{\infty} \chi(y)d(\nu_1-\nu_2)\left(\frac{y}{\phi(\delta)}\right)\right|.$$

The measure $\nu(ax)$ is defined by

$$\nu(ax)(C) = \nu(aC),$$

for C a Borel set in $\mathbf{R}$. Since the integrand of (7.20) vanishes only at $x = 0$, it follows that for fixed δ, ρ^L and $\rho_{\phi,\delta}$ are equivalent. We introduce the strengthened metric

$$\rho_\phi(\nu_1,\nu_2) = \sup_{0<\delta<1} \rho_{\phi,\delta}(\nu_1,\nu_2). \tag{7.21}$$

This will be useful for our integration theory, in which tail behavior of probability distributions will be crucial. Note that infinite distances under p_ϕ are not excluded; this is clearly an inconsequential deviation from the standard properties of a metric. We will now require a proposition. We define for $f\colon \mathbf{R} \to \mathbf{R}$, $\epsilon > 0$,

$$(\mathrm{lip}^{(\epsilon)} f)(x) = \sup_{0<\delta\leq\epsilon} \left|\frac{f(x+\delta)-f(x-\delta)}{\delta}\right|.$$

If ν is a measure on $\mathbf{R}$, then $|\nu|$ is the total mass of ν.

PROPOSITION 7.4: *Let $\beta\colon \mathbf{R} \to \mathbf{R}$ be absolutely continuous and of bounded variation, and ν_1, ν_2 be finite Borel measures on $\mathbf{R}$. Let $\epsilon = \rho^L(\nu_1,\nu_2)$, and assume that $\|\beta\|_\infty$, $\|\beta'\|_1$, and $\|\mathrm{lip}^{(\epsilon)}|\beta'|(x)\|_1$ are finite, where β' denotes the derivative. Then*

$$\int_{-\infty}^{\infty} \beta\, d(\nu_1-\nu_2) \leq \epsilon\Big[\|\beta\|_\infty + 2\|\beta'\|_1 + 2\|\,\mathrm{lip}^{(\epsilon)}|\beta'|(x)\|_1|\nu_2|\Big].$$

Proof: If $\nu_i(x) = \int_{-\infty}^{x} d\nu_i(x')$ (allowing a slight abuse of notation), then

$$\left|\int_{-\infty}^{\infty} \beta\, d(\nu_1-\nu_2)\right| \leq |\beta(x)(\nu_1-\nu_2)(x)|\ \Big|_{-\infty}^{\infty} + \left|\int_{-\infty}^{\infty}(\nu_1-\nu_2)(x)\, d\beta(x)\right|$$

$$\leq \epsilon\|\beta\|_\infty + \int_{-\infty}^{\infty}(\nu_2(x+\epsilon)+\epsilon-(\nu_2(x-\epsilon)-\epsilon))\ |d\beta(x)|$$

$$\leq \epsilon(\|\beta\|_\infty + 2\|\beta'\|_1) + \int_{-\infty}^{\infty}(\nu_2(x+\epsilon)-\nu_2(x-\epsilon))|\beta'(x)|\, dx$$

$$= \epsilon(\|\beta\|_\infty + 2\|\beta'\|_1) + \int_{-\infty}^{\infty}\nu_2(x)(|\beta'(x-\epsilon)| - |\beta'(x+\epsilon)|)\, dx$$

$$= \epsilon(\|\beta\|_\infty + 2\|\beta'\|_1) + \int_{-\infty}^{\infty}\nu_2(x)\, d\eta_\epsilon(x),$$

where

$$\eta_\epsilon(x) = \int_0^x (|\beta'(y-\epsilon)| - |\beta'(y+\epsilon)|)\,dy,$$

and

$$|\eta_\epsilon(x)| \le \epsilon \|\operatorname{lip}^{(\epsilon)}\beta'(x)\|_1.$$

Thus

$$\begin{aligned}\left\| \int_{-\infty}^{\infty} \beta d(\nu_1 - \nu_2) \right\| &\le \epsilon(\|\beta\|_\infty + 2\|\beta'\|_1) + \eta_\epsilon \nu_2 \,|_{-\infty}^{\infty} - \int_{-\infty}^{\infty} \eta_\epsilon \, d\nu_2 \\ &\le \epsilon\Big(\|\beta\|_\infty + 2\|\beta'\|_1 + 2\|\operatorname{lip}^{(\epsilon)} \beta'(x)\|_1 |\nu_2| \Big),\end{aligned}$$

as desired.∎

Remark: If $\overline{\chi}$ is any function for which $f(x) = \frac{\chi(x)-\overline{\chi}(x)}{\theta(x)}$ satisfies the hypotheses of Proposition 7.4, then a metric equivalent to ρ_ϕ obtains by replacing $\overline{\chi}$ by χ in the definition. To see this, note that by Proposition 7.4

$$\left| \frac{1}{\delta} \int_{-\infty}^{\infty} (\chi - \overline{\chi}) d(\nu_1 - \nu_2)\left(\frac{x}{\phi}\right) \right| = \left| \frac{1}{\delta} \int_{-\infty}^{\infty} f\,\theta d(\nu_1 - \nu_2)\left(\frac{x}{\phi}\right) \right| \le F(\rho_\phi(\nu_1, \nu_2)),$$

where $f = \frac{\chi - \overline{\chi}}{\theta}$, and F is a continuous increasing function vanishing at 0.

Let $(\Lambda, \mathcal{B}, \mu)$ be a σ-finite measure space, and $X\colon \Lambda \to \mathcal{D}$ be *measurable,* in that $X^{-1}(\mathcal{O}) \in \mathcal{B}$ for every ρ_ϕ-open $\mathcal{O} \subset \mathcal{D}$. Let $\mathcal{R}(X)$ be the range, and $\mathcal{R}_{\text{ess}}(X) = \{\nu \in \mathcal{R}(X) : \mu(X^{-1}(B_\epsilon(\nu))) > 0 \ \forall \epsilon > 0\}$, where B_ϵ is a ρ_ϕ-ball of radius ϵ. Note for future reference that $\mathcal{R}_{\text{ess}}(X)$ gives the full picture with regard to integration theory, since

$$\mu\{\lambda : X(\lambda) \notin \mathcal{R}_{\text{ess}}(X)\} = 0. \tag{7.22}$$

For otherwise, there would exist an uncountable number of open balls $\{B_m\}_{m \in M}$ in $\mathcal{D}$, with

$$\mu X^{-1}(B_m) = 0, \quad 0 < \mu X^{-1}\left(\bigcup_m B_m \right) < \infty$$

(recall Λ is σ-finite). Such a situation is equivalent (by collapsing each B_i) to a discrete non-atomic measure space of positive finite total measure, which does not exist.

By analogy with the Lebesgue integral, we initially assume μ is finite and X is ρ_ϕ-*bounded,* i.e., the diameter of its range is bounded. Let $\{P_\alpha\}_{\alpha=1}^{\infty}$ be a sequence of at most countable partitions of $\mathcal{R}_{\text{ess}}(X)$, each with elements $\{P_{\alpha i}\}$, $\bigcup_i P_{\alpha i} = \mathcal{R}_{\text{ess}}(X)$. For $S \subset \mathcal{D}$, let $\operatorname{diam} S = \sup\{\rho(\nu_1, \nu_2) : \ \nu_i \in S\}$. We assume

$$(i)\ M(P_\alpha) \equiv \sup_i (\operatorname{diam} P_{\alpha i}) \underset{\alpha \to \infty}{\longrightarrow} 0$$

$$(ii)\ m(P_\alpha) \equiv \sup_i \mu(P_{\alpha i}) \underset{\alpha \to \infty}{\longrightarrow} 0.$$

Hence the mesh of P_α vanishes both in diameter (*i*) and measure (*ii*). For the purposes of this definition, as in Chapter 3, atoms p of μ may be artificially divided into pieces p_j and apportioned to various $P_{\alpha i}$, as long as $\sum \mu(p_j) = \mu(p)$.

DEFINITION 7.5: A partition net satisfying (*i*) and (*ii*) is *infinitesimal.*

Let $\phi(\delta)$ be defined for small positive δ, and vanish at 0. We say that

$$_L\int_\Lambda X(\lambda)\phi(d\mu(\lambda)) \equiv \lim_{\alpha\to\infty} \sum X(\lambda_{\alpha i})\phi(\mu X^{-1}(P_{\alpha i}))$$

exists if the right side is independent of choice of $\lambda_{\alpha i} \in P_{\alpha i}$ and of $\{P_\alpha\}_{i=1}^\infty$ satisfying (*i*) and (*ii*). This is the *finite Lebesgue integral* of X with respect to ϕ.

Remark: The finite Lebesgue integral does not depend on whether the partitions $P_{\alpha i}$ are over the range $\mathcal{R}(X)$ or the essential range $\mathcal{R}_{\text{ess}}(X)$.

By our assumption of σ-finiteness of Λ, $\mathcal{R}_{\text{ess}}(X)$ is separable. For if it were not, there would exist an uncountable disjoint collection $\{B_\alpha\}$ of balls in $\mathcal{D}$, with $\mu(X^{-1}(B_\alpha)) > 0$, contradicting the σ-finiteness of μ.

For the case of a general σ-finite measure μ and measurable function X: $\Lambda \to \mathcal{D}$, we disjointly partition $\mathcal{R}_{\text{ess}} = \bigcup_{P_i\in P} P_i$; we assume $\mu^*(P_i)$, diam $P_i < \infty$, where $\mu^* = \mu X^{-1}$. The partition is always assumed at most countable, which is possible since $\mathcal{R}_{\text{ess}}(X)$ is separable and σ-finite. The general Lebesgue integral

$$_L\int_\Lambda X(\lambda)\phi(d\mu(\lambda)) \equiv \int_{\mathcal{R}_{\text{ess}}} \nu\phi(d\mu^*(\nu)) \equiv \sum_i{}^* \; _L\int_{X^{-1}(P_i)} X\phi(d\mu) \tag{7.23}$$

is defined using finite integrals on the right, and exists if the sum on the right is independent of the partition P. The above independence condition, which may seem difficult to test, is natural; see Theorem 8.4.

We must verify that the general Lebesgue integral coincides with the finite one if $\mathcal{R}(X)$ is finite, bounded and separable. To this end, we require

PROPOSITION 7.6: *If $\mathcal{D}_1 \subset \mathcal{D}$ and the finite Lebesgue integral $_L\int_{\mathcal{D}_1} \nu\phi(d\mu^*)$ exists, and if $\mathcal{D}_2 \subset \mathcal{D}_1$ has positive Borel measure, then $\int_{\mathcal{D}_2} \nu\phi(d\mu^*)$ exists.*

PROPOSITION 7.7: *If the finite integral $_L\int_{\mathcal{D}_1} \nu\phi(d\mu^*)$ exists and P_α is a partition of $\mathcal{D}_1$, then*

$$_L\int_{\mathcal{D}_1} \nu\phi(d\mu^*(\nu)) = \sum_k{}^* \; _L\int_{P_{\alpha k}} \nu\,\phi(d\mu^*(\nu)), \tag{7.24}$$

order independently, where $\sum_k^* \nu_k = \nu_1 * \nu_2 * \ldots$.

Proof of Proposition 7.6: We will require the fact that the finite Lebesgue integral is always infinitely divisible (this follows from the definition and a stronger result is proved in Theorem 7.9). Hence the characteristic function $\Phi(t)$ of $Y = \int_{D_1} \nu\, \phi(d\mu^*)$ does not vanish. Let $P_\alpha^{(2)}$ be an infinitesimal sequence of partitions of D_2, and $P_\alpha^{(1)}$ a sequence for $D_1 \sim D_2$. Let $X_{\alpha i}^{(1)} \in P_{\alpha i}^{(1)}$, $X_{\alpha i}^{(2)} \in P_{\alpha i}^{(2)}$, and

$$X_\alpha^j = \sum_i^* X_{\alpha i}^{(j)} \phi(\mu_{\alpha i}^{(j)}) \qquad (j = 1, 2).$$

If $X_\alpha^{(j)}$ is unbounded (i.e., has a subsequence converging in law to a measure strictly less than 1 as $\alpha \to \infty$) for $j = 1$ or 2, then Y cannot exist. Thus, there is a subsequence of $\{X_\alpha^{(1)}\}$ which converges in law to some probability distribution, and without loss of generality we reindex $X_\alpha^{(1)}$ so that $X_\alpha^{(1)} \underset{\alpha\to\infty}{\Rightarrow} X^{(1)}$, where $X^{(1)}$ has ch.f. $\Phi^{(1)}$. Thus,

$$X_\alpha^{(1)} * X_\alpha^{(2)} \Rightarrow Y\,;$$

hence

$$X^{(1)} * X_\alpha^{(2)} \Rightarrow Y\ ,$$

or in terms of characteristic functions,

$$\Phi^{(1)}(t)\Phi_\alpha^{(2)}(t) \underset{\alpha\to\infty}{\to} u(t). \tag{7.25}$$

Since $\Phi(t) \neq 0$, this shows that $\Phi_\alpha^{(2)}(t)$ converges, as does $X_\alpha^{(2)}$ in law. By (7.25), w$-\lim_{\alpha\to\infty} X_\alpha^{(2)}$ is independent of the choice $P_\alpha^{(2)}$, proving Proposition 7.6.∎.

Sketch of proof of Proposition 7.7: Note that terms on the right in (7.24) can be well approximated by "Lebesgue sums" $\sum X_{\alpha' i}\phi(\mu_{\alpha' i}^*)$, where $\{X_{\alpha' i}\}$ are elements of a subpartition of P_α. Since the approximants can be made to converge order independently to the left side, so can their limits $\int_{P_{\alpha k}} \nu\, \phi(d\mu^*(\nu))$. ∎.

In this chapter an assortment of distributions will obtain as Lebesgue integrals, as no prior constraints are placed on existence of the mean and variance of the integrand.

The Lebesgue integral is clearly linear, i.e.,

$${}_L\!\int (X_1 + X_2)\, \phi(d\mu^*) = {}_L\!\int X_1 \phi(d\mu^*) + {}_L\!\int X_2 \phi(d\mu^*),$$

if X_1 and X_2 are independent for each $\lambda \in \Lambda$. It is also additive, i.e., the integral over a union $E_1 \cup E_2$ of disjoint sets is the sum (i.e., convolution, in the distribution picture) of the integrals over E_1 and E_2.

Henceforth, all integrals of real-valued functions will be Lebesgue integrals. Clearly, our integral reduces to standard Lebesgue integration when ϕ is the identity, and $X(\lambda)$ is a point mass for $\lambda \in \Lambda$.

THEOREM 7.8: *Let $X(\lambda)$ have 0 mean a.e. $[\mu]$, with $v = \int_\Lambda \mathcal{E}(X^2)\,d\mu$ and $\int_\Lambda |\mathcal{E}(X^3)|\,d\mu$ finite. Then the Lebesgue integral of X exists, and*

$$ {}_L\int_\Lambda X(\lambda)(d\mu(\lambda))^{\frac{1}{2}} = N(0, v) $$

Proof: Let $\{P_\alpha\}$ and $\{\lambda_{\alpha i}\}$ be as above, and let $X_{\alpha i} = X(\lambda_{\alpha i})$, $\mu_{\alpha i} = \mu X^{-1}(P_{\alpha i})$. We invoke Theorem 3.6, and note that

$$ \frac{\gamma_\alpha^3}{\sigma_\alpha^3} \equiv \frac{\sum_i \mathcal{E}(|X_{\alpha i}|^3)\mu_{\alpha i}^{\frac{3}{2}}}{(\sum_i \mathcal{E}(X_{\alpha i}^2)\mu_{\alpha i})^{\frac{3}{2}}} \underset{\alpha\to 0}{\longrightarrow} 0, $$

since the denominator approaches the integral $\left(\int \mathcal{E}(X^2(\lambda))d\mu(\lambda)\right)^{\frac{3}{2}}$, while the numerator vanishes, given that $\sup_i \mu_{\alpha i} \underset{\alpha\to\infty}{\longrightarrow} 0$. This shows

$$ \sum_i X_{\alpha i}\mu_{\alpha i}^{\frac{1}{2}} \underset{\alpha\to\infty}{\Rightarrow} N(0, v). \blacksquare $$

§ 7.3 Basic Properties

DEFINITION 7.9: A probability distribution ν is *stable* if to every $a_1, a_2 > 0$, b_1, b_2, there correspond constants $a > 0$ and b such that

$$ F(a_1x + b_1) * F(a_2x + b_2) = F(ax + b), $$

where F is the d.f. of ν. If b_1, b_2, b can be chosen to be 0, then ν is a *scaling stable*.

Stable distributions, which are intimately connected to the Lebesgue integral, can be characterized by

THEOREM 7.10 (Khinchin and Lévy, [KL]): *In order that ν be stable, it is necessary and sufficient that its characteristic function Φ satisfy*

$$ \psi(t) \equiv \ln \Phi(t) = i\gamma t - c|t|^\eta\left\{1 + i\beta\frac{t}{|t|}\omega(t,\eta)\right\}, \tag{7.26} $$

where $\gamma \in \mathbf{R}$, $-1 \leq \beta \leq 1$, $0 < \eta \leq 2$, $c \geq 0$, and

$$\omega(t,\eta) = \begin{cases} \tan \frac{\pi}{2}\eta & \text{if } \eta \neq 1 \\ \frac{2}{\pi}\ln|t| & \text{if } \eta = 1 \end{cases}.$$

COROLLARY 7.10.1: *In order that ν be scaling stable, it is necessary and sufficient that the logarithm of its characteristic function have the form*

$$\psi(t) = -c_1|t|^\eta + ic_2|t|^\eta \frac{t}{|t|}, \tag{7.27}$$

where $c_1 \geq 0$, $0 < \eta \leq 2$, and $|c_2| \leq c_1 |\tan \frac{\pi}{2}\eta|$; if $\eta = 1, c_2 \in \mathbf{R}$ is arbitrary.

Proof: The scaling stability condition is $\psi(\frac{t}{a_1}) + \psi(\frac{t}{a_2}) = \psi(\frac{t}{a})$. Sufficiency is clear, so we prove only necessity. If ν is stable, $\psi(t)$ is given by eq. (7.26). We assume momentarily that $\eta \neq 1$. Using (7.26), we have

$$i\gamma t\left(\frac{1}{a_1} + \frac{1}{a_2}\right) - c|t|^\eta\left(\frac{1}{a_1^\eta} + \frac{1}{a_2^\eta}\right)\left\{1 + i\beta\frac{t}{|t|}\tan\frac{\pi}{2}\eta\right\} =$$

$$i\gamma t\left(\frac{1}{a}\right) - c|t|^\eta\left(\frac{1}{a^\eta}\right)\left\{1 + i\beta\frac{t}{|t|}\tan\frac{\pi}{2}\eta\right\}.$$

We conclude that (i) $\gamma = 0$ or (ii) $c = 0$. In either case, the function ψ fits the form (7.26). If $\eta = 1$, a similar argument shows $\beta = 0$, completing the proof. ∎

The proof of the following lemma uses arguments similar to those in Theorem 7.3 and its Corollary, and is omitted.

LEMMA 7.11.1: *Let $\{X_i\}_i$ be independent r.v.'s, and $\sum_i X_i$ converge order independently. Let $\{a_{ni}\}$ be an infinitesimal array of real numbers (i.e., $a_{ni} \underset{n\to\infty}{\to} 0$, uniformly in i). Then $\sum_i a_{ni}X_i \underset{n\to\infty}{\Rightarrow} 0$.*

We have the following characterization of Lebesgue integrals.

THEOREM 7.11: *In order that ν be the distribution of a Lebesgue integral of an r.v.-valued function, it is necessary and sufficient that ν be scaling stable.*

Proof: Assume $Y = {}_L\int X(\lambda)\phi(du(\lambda)) \neq 0$. It is easy to see that then $\phi(\delta) \underset{\delta\to 0}{\to} 0$. Let $\mathcal{D}_1$ be the essential range of X and μ^* the induced measure on $\mathcal{D}_1$. Since scaling stability is preserved under sums, there is no loss in assuming $\mathcal{D}_1$ is ρ_ϕ-bounded and finite in measure. Let $0 = r_0 < r_1 < r_2 < \ldots$ be a sequence of positive numbers. Let $P^k_{r_k}$ be a partition of $\mathcal{D}_1$

into an at most countable collection of sets of measure smaller than e^{-r_k}, with ρ_ϕ-diameter less than $\frac{1}{k}$. For $r \geq r_k$, let P_r^k be a sub-partition of $P_{r_k}^k$ such that given any $P_{r_k j}^k \in P_{r_k}^k$, there is at most one element $P_{rj_1}^k \in P_r^k$ with $P_{rj_1}^k \subset P_{r_k j}^k$ and $\mu^*(P_{rj_1}^k) \neq e^{-r}$ (note j_1 changes with r and j). For simplicity, we assume such an element exists for all $r > r_k$; its measure may be 0. Recall that singletons $\nu \in P_{r_k}$ can be sub-divided (along with their measures), for purposes of partitioning; if this were only allowed for singletons of positive measure, the following arguments would be somewhat more technical. Fix j as above, and let P_r^k be constructed in such a way that $\left(\bigcap_{r_k \leq r < r_{k+1}} P_{rj_1}^k\right) \bigcap P_{r_k j}^k$ is non-empty, containing a fixed element, denoted by $X_{r_k j_1}^k$, for each j. For $r \geq r_k$, choose $X_{rj_1}^k = X_{r_k j_1}^k \in P_{rj_1}^k$, where $X_{r_k j_1}^k$ is as above.

For each $P_{rj}^k \in P_r^k$ with $\mu(P_{rj}^k) = e^{-r}$, choose $X_{rj}^k \in P_{rj}^k$. For $r \geq r_k$, let $P_r^{k*} = \{P_{rj}^k \in P_r : \mu^*(P_{rj}^k) \neq e^{-r}\}$, and consider

$$S_r^{k*} = \sum_{P_{rj} \in P_r^{k*}} X_{rj}^k \phi(\mu^*(P_{rj}^k)), \tag{7.28}$$

which is order-independent by definition of the Lebesgue integral. Given $r_0, \ldots, r_k$, we note that since $\mu^*(P_{rj}) \leq e^{-r}$, r_{k+1} can be made sufficiently large that for $r \geq r_{k+1}$, S_r^{k*} is arbitrarily small (i.e., close to the unit mass at 0). This follows by Lemma 7.11.1, and from the fact that at most are representative of each of the sets $\{P_{r_k j}^k\}_j$ appears in the sum (7.28). We thus successively select $r_0, r_1, \ldots$ so that $\rho^L(S_r^{k*}, \delta_0) \underset{k\to\infty}{\to} 0$, uniformly in $r \in [r_{k+1}, \infty)$.

We now choose our final partitions. Let

$$P_r \equiv P_r^k, \quad X_{rj} \equiv X_{rj}^k, \quad P_r^* \equiv P_r^{k*}, \quad S_r^* \equiv S_r^{*k} \qquad (r \in [r_{k+1}, r_{k+2})).$$

By the above,

$$S_r^* \underset{r\to\infty}{\Rightarrow} 0.$$

Thus,

$$S_r \equiv \sum_{P_{rj} \notin P_r^*} X_{rj} \phi(e^{-r}) \underset{r\to\infty}{\Rightarrow} Y = {}_L\int_\Lambda X(\lambda)\phi(d\mu(\lambda)). \tag{7.29}$$

However, each element of $P_r \sim P_r^*$ may be divided exactly in half, $P_{rj} = P_{rj}^{(1)} \cup P_{rj}^{(2)}$, and we may choose independent copies $X_{rj}^{(1)}$, $X_{rj}^{(2)}$ of X_{rj}, to be contained in $P_r^{(1)}$ and $P_r^{(2)}$; respectively. Then

$$\sum_{P_{rj} \notin P_r^*} X_{rj}^{(1)} \phi\left(\frac{e^{-r}}{2}\right) + \sum_{P_{rj} \notin P_r^*} X_{rj}^{(2)} \phi\left(\frac{e^{-r}}{2}\right) = \frac{\phi\left(\frac{e^{-r}}{2}\right)}{\phi(e^{-r})}(S_r * S_r) \underset{r\to\infty}{\Rightarrow} Y.$$

Thus $l(2) = \lim_{\delta\to 0} \dfrac{\phi(2\delta)}{\phi(\delta)}$ exists, and $Y^{*2} \equiv Y * Y = l(2)Y$ in distribution. Similarly, for $n \in \mathbf{N}$, $Y^{*n} = l(n)Y$ for some $l(n) \in \mathbf{R}$. Thus Y is infinitely divisible; if ψ is the logarithm of its ch.f. then

$$n\psi(t) = \psi(l(n)t) \qquad (n = 1, 2, \ldots). \tag{7.30a}$$

Thus,

$$\frac{1}{m}\psi(t) = \psi\left(\frac{1}{l(m)}t\right), \tag{7.30b}$$

and $\frac{n}{m}\psi(t) = \psi(\frac{l(n)}{l(m)}t)$. We define $l(\frac{n}{m}) = \frac{l(n)}{l(m)}$. If $\lambda \geq 0$, let $q_i \to \lambda$, $q_i \in \mathbf{Q}$, the rationals. Then

$$\lambda\psi(t) = \lim_{i\to\infty} q_i\psi(t) = \lim_{i\to\infty} \psi(l(q_i)t).$$

This holds for all $t \in \mathbf{R}$. Thus if ψ is not identically 0, $\lim_{i\to\infty} l(q_i) \equiv l(\lambda)$ exists, l is continuous, and $\lambda\psi(t) = \psi(l(\lambda)t)$ $(\lambda \in \mathbf{R}^+)$. Letting $q_i \to 0$, we have $l(0) = 0$. Thus, if $r > 0$, $\frac{l(\lambda_1)}{l(\lambda_2)} = r$ for some $\lambda_1, \lambda_2 > 0$. Thus, if $a_1, a_2 > 0$, then letting $r = \frac{a_1}{a_2}$, there exists $c > 0$ such that

$$\begin{aligned}
\psi(a_1 t) + \psi(a_2 t) &= \psi(cl(\lambda_1)t) + \psi(cl(\lambda_2)t) \\
&= (\lambda_1 + \lambda_2)\psi(ct) \qquad\qquad (7.31) \\
&= \psi(l(\lambda_1 + \lambda_2)ct),
\end{aligned}$$

so that Y is scaling stable.

Conversely, assume Y is a scaling stable r.v. Let (7.27) be its log ch.f. and $\Lambda = \{Y\}$ be a singleton space with measure one. For each $\alpha \in \mathbf{N}$, let $\{Y_{\alpha i}\}_i$ be independent copies of Y in Λ which are assigned measure $\mu_{i\alpha}$, and let $\phi(\lambda) = \lambda^\eta$. Then $S_\alpha = \sum_i Y_{\alpha i}(\mu_{\alpha i})^{\frac{1}{\eta}}$ has ch.f.

$$\begin{aligned}
\psi_\alpha(t) &= \sum_i -c_1\left|\mu_{i\alpha}^{\frac{1}{\eta}}t\right|^\eta + ic_2\left|\mu_{i\alpha}^{\frac{1}{\eta}}t\right|^\eta \frac{t}{|t|} \\
&= \left(\sum_i \mu_{i\alpha}\right)\left(-c_1|t|^\eta + ic_2|t|^\eta\frac{t}{|t|}\right), \qquad\qquad (7.32) \\
&= -c_1|t|^\eta + ic_2|t|^\eta\frac{t}{|t|},
\end{aligned}$$

so that the distribution of S_α is independent of α. Letting Y also denote the identity function on Λ, this shows that $\int_\Lambda Y\phi(d\mu) = \lim_{\alpha\to\infty} S_\alpha = Y$ in distribution.∎

We now show that the scaling function ϕ must be very restricted for a (non-trivial) Lebesgue integral to exist. To properly motivate this, we make some observations about so-called semi-stable stochastic processes [La]. A stochastic process X_t on $\mathbf{R}$ is *semi-stable* if for every $a > 0$, $X_{at} \simeq b(a)X_t + c(a)$ with $b, c \in \mathbf{R}$, and $\simeq$ denoting isomorphism. We assume X_t is continuous, i.e., that

$$\lim_{h\to 0} P(|X_{t+h} - X_t| > \epsilon) = 0,$$

and that X_t is proper, i.e. non-degenerate for all t. We then have

THEOREM 7.12 [La]: *If X_t is semi-stable, proper, and continuous in the above sense, and if X_t has independent increments, then*

$$b(a) = a^{\alpha},$$

for some $\alpha \geq 0$.

With the proper interpretation, this may be viewed as a special case of the next theorem. The identification depends on the fact that any semi-stable, stationary 0-mean stochastic process Y_t with independent increments can be written

$$Y_t = \int_{-\infty}^{t} X(t)\ \phi(dt),$$

where X is a measurable r.v.-valued function and ϕ: $\mathbf{R}^+ \to \mathbf{R}^+$.

THEOREM 7.13: *If $Y = {}_L\int_\Lambda X\phi(d\mu)$ exists and is non-zero, then there exists $\frac{1}{\eta} > 0$ such that $\frac{\phi(\delta)}{\delta^{\frac{1}{\eta}}} \underset{\delta\to 0}{\to} c$ for some $c > 0$. Specifically, η is given by (7.27), where ψ is the log ch.f. of Y.*

Proof: Note that convergence of Y implies $\phi(\delta) \underset{\delta\to 0}{\to} 0$. Let $\mathcal{D}_1 = \mathcal{R}_{\mathrm{ess}}(X)$. Since $Y = \int_{\mathcal{D}_1} \nu\phi(d\mu^*)$ exists ($\mu^* = \mu X^{-1}$), it does also (and is non-zero) on some bounded, finite sub-domain of $\mathcal{D}_1$. Thus assume without loss that $\mathcal{D}_1$ is ρ_ϕ-bounded and μ^*-finite. Let S_r be constructed as in (7.29). Let each P_{rj} contributing to S_r be subdivided into n equal pieces, each containing X_{rj} (again, arguments become technical if formal subdivision of X_{rj} is not allowed). This subdivision gives

$$\left(\sum_{P_{rj}\notin P_r^*} X_{rj}\phi\left(\frac{e^{-r}}{n}\right)\right)^{*n} = \left(\sum_{P_{rj}\notin P_r^*} X_{rj}\phi(e^{-r})\right)^{*n} \frac{\phi\left(\frac{e^{-r}}{n}\right)}{\phi(e^{-r})} \underset{r\to\infty}{\Rightarrow} Y, \tag{7.33}$$

so that, letting $r \to \infty$,

$$Y = \frac{Y^{*n}}{l(n)}$$

in distribution, where $l(n) = \lim \frac{\phi(n\delta)}{\phi(\delta)}$. By (7.27), therefore, $l(n) = n^{\frac{1}{\eta}}$. Thus, $\phi(\frac{1}{n}) \sim cn^{-\frac{1}{\eta}}$ for some $c > 0$.

We now assume $\phi(\delta)$ is not asymptotic to $c\delta^{\frac{1}{\eta}}$, for a contradiction. Assume without loss that for some sequence $\delta_n \to 0$, $\phi(\delta_n) \geq (c+\epsilon)\delta_n^{\frac{1}{\eta}}$, for $\epsilon > 0$. Let $\delta_n = e^{-s_n}$. Then by the proof of Theorem 7.11,

$$S_{s_n} = \phi(\delta_n) \sum_{P_{s_n j} \notin P^*_{s_n}} X_{s_n j} \underset{n\to\infty}{\Rightarrow} Y. \tag{7.34}$$

We assume without loss that the numbers r_n (see proof of Theorem 7.11) have the property that e^{r_n} is integral.

Let $\bar{s}_n$ be defined by $e^{\bar{s}_n} = [e^{s_n}]$, where $[\cdot]$ denotes greatest integer. By readjustment of $\{r_n\}$ in the proof of Theorem 7.11, we assume without loss that s_n and $\bar{s}_n$ lie in a single interval $[r_k, r_{k+1})$. Furthermore, we can choose P_r such that $\{X_{sj}$: X_{sj} appears in $S_s\} \supset \{X_{\bar{s}_n j}$: $X_{\bar{s}_n j}$ appears in $S_{\bar{s}_n}\}$ for $\bar{s}_n \leq s \leq \bar{s}_{n+1}$. If $\tilde{X}_n = \{X_{sj}$: X_{sj} appears in S_s for some $\bar{s}_n \leq s \leq \bar{s}_{n+1}$, but not in $S_{\bar{s}_n}\}$ (we treat all distinct elements of $\tilde{X}_n$ as independent), then

$$\tilde{S}_n = \sum_{X \in \tilde{X}_n} X\phi(e^{-s_n}) \underset{n\to\infty}{\Rightarrow} 0. \tag{7.35}$$

Since the argument for this is similar to one for S_n^* in the proof of the last theorem, we omit it.

It follows from (7.34) and (7.35) that

$$\phi(e^{-s_n}) \sum_{P_{\bar{s}_n j} \notin P^*_{\bar{s}_n}} X_{\bar{s}_n j} \underset{n\to\infty}{\Rightarrow} Y.$$

This provides the contradiction, since

$$\phi(e^{-\bar{s}_n}) \sum_{P_{\bar{s}_n j} \notin P^*_{\bar{s}_n}} X_{\bar{s}_n j} \underset{n\to\infty}{\Rightarrow} Y, \tag{7.36}$$

and by assumption $\frac{\phi(e^{-s_n})}{\phi(e^{-\bar{s}_n})}$ fails to converge to 1 as $n \to \infty$. ∎.

It follows from Theorem 7.13 and Corollary 7.3.1 that the set of admissible functions ϕ (i.e., measurable functions yielding non-trivial integrals) fall into equivalence classes with $\phi_1 \sim \phi_2$ if $\frac{\phi_1(\delta)}{\phi_2(\delta)} \underset{\delta\to 0}{\to} c$ for some constant c. Furthermore, each class has exactly one homogeneous representative $\phi(\delta) = \delta^{\frac{1}{\eta}}$. Thus the classes of ϕ are indexed by the positive reals.

CHAPTER 8

INTEGRABILITY CRITERIA AND SOME APPLICATIONS

In the first half of this chapter, we introduce an important integrability criterion for random variable-valued functions (Theorem 1). We then compare the Lebesgue with the Riemann integral, and finally give applications to the calculation of asymptotic joint distributions of commuting observables.

A central result shows the non-linear Lebesgue integral of Chapter 7 to be continuously imbeddable into a standard Lebesgue integral I^* over the space M^* of measures on the Borel sets of $\mathcal{D}$, the space of probability distributions. Let $\mathcal{C}$ be the space of logarithms of ch.f.'s (log ch.f.'s) of infinitely divisible distributions. Let $J : \mathcal{D} \to \mathcal{C}$ be the partially defined function $J\nu = \lim_{\delta \to 0} \int (e^{itx} - 1) d\nu_\delta$, where $\nu_\delta = \frac{1}{\delta}\nu\left(\frac{x}{\phi(\delta)}\right)$. Let μ be a measure on the Borel sets of $\mathcal{D}$, and $I^* : M^* \to \mathcal{C}$ denote the integration operator

$$I^*\mu = \int J\nu \, d\mu(\nu).$$

Finally, let $\mathcal{F}$ denote the operation taking a ch.f. to its probability distribution, and E : $\mathcal{C} \to \mathcal{D}$ be given by $E\psi = \mathcal{F}(e^\psi)$.

Theorem 8.5 states that if $I_\phi : M^* \to \mathcal{D}$ denotes the Lebesgue integral

$$I_\phi\mu \;=\; {}_L\int_{\mathcal{D}} \nu \, \phi(d\mu\ (\nu)),$$

then the diagram

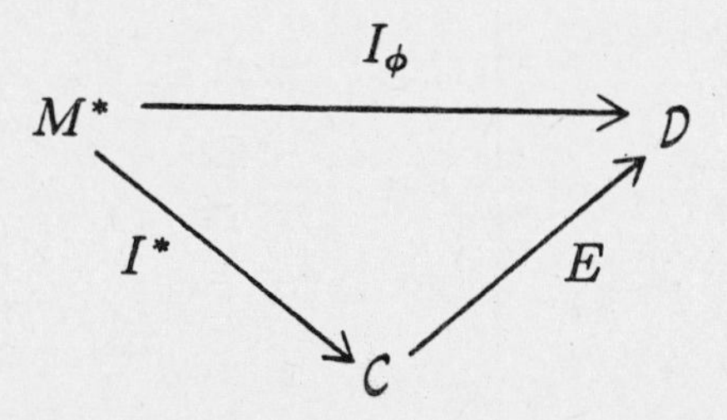

commutes. Specifically, the Lebesgue integral of a r.v.-valued function $X : \Lambda \to \mathcal{D}$ is

$${}_L\int_\Lambda X\phi(d\mu) = \mathcal{F}\left(\exp \int_\Lambda \lim_{\delta \to 0} \frac{1}{\delta}\left(\mathcal{E}\left(e^{iX(\lambda)t\phi(\delta)}\right) - 1\right) d\mu(\lambda)\right).$$

§8.1 The Criterion

The metric ρ_ϕ (eq. (7.21)) on the space $\mathcal{D}$ of probability laws will be used throughout this chapter. Let $X : \Lambda \to \mathcal{D}_1 \subset \mathcal{D}$ be measurable from a space $(\Lambda, \mathcal{A}, \mu)$ and $\mu^* = \mu X^{-1}$. For measures ν_i, let $\sum_i^* \nu_i = \nu_1 * \nu_2 * \ldots$; as usual, let $\sum_i \nu_i = \nu_1 + \nu_2 + \ldots$.

We say $\mathcal{D}_1$ is *ϕ-separable* if it is ρ_ϕ-separable. Note that ϕ-Lebesgue integrability is senseless if $\mathcal{D}_1$ or some subset containing the essential part of the range is not ϕ-separable, since required partitions P_α cannot be countable. A function $X : \Lambda \to \mathcal{D}$ is *bounded* if its range has finite ρ_ϕ-diameter. We now show that, in complete analogy to the real-valued case, if μ is finite and $X : \Lambda \to \mathcal{D}_1$ is bounded and measurable, then it is Lebesgue integrable. We require

LEMMA 8.1.1:
(a) Let ν_{1i}, ν_{2i} be two countable families of finite Borel measures on $\mathbf{R}$. Then

$$\rho^L\Big(\sum_i a_i\nu_{1i}, \sum_i a_i\nu_{2i}\Big) \le \max\Big(\sum_i a_i\rho^L(\nu_{1i},\nu_{2i}), \sup_i \rho^L(\nu_{1i},\nu_{2i})\Big) \tag{8.1}$$

(b) Similarly, if $(\mathcal{D}, \mathcal{A}, \mu^*)$ is a measure space and $F_\nu(x)$, $G_\nu(x)$ are d.f.'s, then

$$\rho^L\Big(\int_{\mathcal{D}} F_\nu(x)d\mu^*(\nu), \int_{\mathcal{D}} G_\nu(x)d\mu^*(\nu)\Big)$$

$$\le \max\Big(\int_{\mathcal{D}} \rho^L(F_\nu, G_\nu)d\mu^*(\nu), \sup_\nu \rho^L(F_\nu, G_\nu)\Big).$$

Proof of (a): Set $\epsilon_i = \rho^L(\nu_{1i}, \nu_{2i})$, and let ϵ be the right side of (8.1). Letting F_{ni} be the corresponding distribution functions,

$$\begin{aligned}\sum(a_iF_{1i}(x-\epsilon)) - \epsilon \le \sum a_i(F_{1i}(x-\epsilon_i)-\epsilon_i) &\le \sum a_iF_{2i}(x)\\ &\le \sum a_i(F_{1i}(x+\epsilon_i)+\epsilon_i)\\ &\le \sum(a_iF_{1i}(x+\epsilon)) + \epsilon. \blacksquare\end{aligned}$$

For a function X from a measure space Λ to a topological space $\mathcal{D}$ recall that $\nu \in \mathcal{D}$ is in the essential range $\mathcal{R}_{\text{ess}}(X)$ if $\mu X^{-1}(N) > 0$ for every neighborhood N of ν.

Recall also

$$\theta(x) = \frac{x^2}{1+x^2}, \qquad \chi(x) = \frac{x}{1+x^2}, \qquad \chi^*(x) = \begin{cases} x; & |x| \le 1 \\ \frac{1}{x}; & |x| \ge 1 \end{cases}. \tag{8.2}$$

THEOREM 8.1: *Let $(\Lambda, \mathcal{B}, \mu)$ be finite, and $X : \Lambda \to \mathcal{D}$, be bounded and ρ_ϕ-Borel measurable. Then $Y = {}_L\int_\Lambda X(\lambda)\phi(d\mu(\lambda))$ exists if and only if*

$$G(\nu; x) \equiv \text{w-}\lim_{\delta\to 0} \int_{-\infty}^{x} \frac{y^2}{1+y^2} d\nu_\delta(y), \qquad \gamma_\nu \equiv \lim_{\delta\to 0} \int_{-\infty}^{\infty} \frac{y}{1+y^2} d\nu_\delta(y) \tag{8.3}$$

exist for every ν in the essential range of X, where $\nu_\delta = \frac{1}{\delta}\nu\left(\frac{x}{\phi(\delta)}\right)$. Furthermore, the pair

$$\gamma = \int \gamma_X \mu(\lambda), \quad G(x) = \int G(X; x) d\mu(\lambda).$$

is the Lévy-Khinchin transform of Y.

Proof: Assume $\gamma(\nu)$ and $G(\nu)$ exist for $\nu \in \operatorname{supp} \mu^*$ where $\mu^* = \mu X^{-1}$. Let $\{P_\alpha\}_{\alpha=1}^{\infty}$ be an infinitesimal sequence of partitions of $\mathcal{D}_1 = \mathcal{R}_{\text{ess}}(X)$. For $\nu_{\alpha k} \in P_{\alpha k}$, we first verify that the distributions $\frac{1}{\phi(\mu_{\alpha k})}\nu_{\alpha k}\left(\frac{x}{\phi(\mu_{\alpha k})}\right)$ form an infinitesimal sequence. To this end, we need to show $\frac{1}{\phi(\delta)}\nu\left(\frac{x}{\phi(\delta)}\right) \underset{\delta\to\infty}{\Rightarrow} \delta_0$ uniformly in $\nu \in \mathcal{R}_{\text{ess}}(X)$. Suppose this is false. Then there exists $\epsilon > 0$ such that for every $N > 0$, there is a $\nu \in \mathcal{R}_{\text{ess}}(X)$ such that $\nu(-\infty, -N) + \nu(N, \infty) \geq \epsilon$. This contradicts the boundedness (i.e., finite diameter) of $\mathcal{R}_{\text{ess}}(X)$, since then $\sup_\delta \frac{1}{\delta} \int_{-\infty}^{\infty} \theta(y) d\nu\left(\frac{x}{\phi(\delta)}\right)$ can be made arbitrarily large for $\nu \in \mathcal{R}_{\text{ess}}(X)$. Thus the sequence is infinitesimal.

According to Corollary 7.3.1, it now suffices to show

$$\sum_k b_{\alpha k} + \int_{-\infty}^{\infty} \chi \, d\nu_{\alpha k}\left(\frac{y + b_{\alpha k}}{\phi_{\alpha k}}\right) \underset{\alpha\to\infty}{\to} \gamma \tag{8.4a}$$

$$\sum_k \int_{-\infty}^{x} \theta \, d\nu_{\alpha k}\left(\frac{y + b_{\alpha k}}{\phi_{\alpha k}}\right) \underset{\alpha\to\infty}{\Rightarrow} G(x), \tag{8.4b}$$

where $\phi_{\alpha k} = \phi(\mu_{\alpha k})$, and

$$b_{\alpha k} = \int_{-\infty}^{\infty} \chi^*(y) d\nu_{\alpha k}\left(\frac{y}{\phi_{\alpha k}}\right). \tag{8.5}$$

First consider (8.4b). Note that

$$\begin{aligned} \sum_k \int_{-\infty}^{x} \theta(y) \, d\nu_{\alpha k}\left(\frac{y + b_{\alpha k}}{\phi(\mu_{\alpha k})}\right) &= \sum_k \int_{-\infty}^{x + b_{\alpha k}} \theta(y) \, d\nu_{\alpha k}\left(\frac{y}{\phi(\mu_{\alpha k})}\right) \\ &\quad - b_{\alpha k} \int_{-\infty}^{x + b_{\alpha k}} \theta'(y + b^*_{\alpha k}) \, d\nu_{\alpha k}\left(\frac{y}{\phi(\mu_{\alpha k})}\right) \end{aligned} \tag{8.6}$$

where $|b^*_{\alpha k}| \leq |b_{\alpha k}|$. In addition, $\frac{1}{\mu_{\alpha k}} b_{\alpha k}$ is bounded uniformly in α and k (for α sufficiently large) for all non-vanishing $\mu_{\alpha k}$, since the support of μ^* is bounded and replacing χ in the metric ρ_ϕ by χ^* gives an equivalent one (see the remark after Prop. 7.4.). The second term in the sum on the right of (8.6) converges to 0 uniformly in x and k since $\left\{\nu_{\alpha k}\left(\frac{x}{\phi(\mu_{\alpha k})}\right)\right\}_k$ are infinitesimal, so that

$$\rho^L\left(\sum_k \int_{-\infty}^{x} \theta\, d\nu_{\alpha k}\left(\frac{y+b_{\alpha k}}{\phi(\mu_{\alpha k})}\right), \sum_k \int_{-\infty}^{x} \theta d\nu_{\alpha k}\left(\frac{y}{\phi(\mu_{\alpha k})}\right)\right) \underset{\alpha\to\infty}{\longrightarrow} 0. \tag{8.7}$$

Let $G_\delta(\nu; x) = \frac{1}{\delta}\int_{-\infty}^{x} \theta(x) d\nu\left(\frac{y}{\phi(\delta)}\right)$ be the d.f. of $\theta d\nu_\delta$. Then

$$\begin{aligned}
&\rho^L\left(\int_{\mathcal{D}_1} G(\nu; x) d\mu^*(\nu), \sum_k \int_{-\infty}^{x} \theta\, d\nu_{\alpha k}\left(\frac{y}{\phi(\mu_{\alpha k})}\right)\right) \\
&\qquad \leq \rho^L\left(\sum_k \int_{P_{\alpha k}} G(\nu; x) d\mu^*(\nu), \sum_k \int_{P_{\alpha k}} G(\nu_{\alpha k}; x) d\mu^*(\nu)\right) \\
&\qquad\quad + \rho^L\left(\sum_k \mu_{\alpha k} G(\nu_{\alpha k}; x), \sum_k \mu_{\alpha k} G_{\mu_{\alpha k}}(\nu_{\alpha k}; x)\right).
\end{aligned} \tag{8.8}$$

To show (8.8) vanishes as $\alpha \to \infty$, we note the first term on the right vanishes by Lemma 8.1.1, if

$$\rho^L(G(\nu; x), G(\nu_{\alpha k}, x)) \underset{\alpha\to\infty}{\longrightarrow} 0$$

for $\nu \in P_{\alpha k}$, uniformly in ν and k; the latter follows from the metric infinitesimality of the partition P_α, and the equivalence of ρ^L and $\rho_{\phi,\delta}$ for fixed $\delta > 0$.

The second term on the right requires some care, however. For $\epsilon, \delta > 0$, let $E_{\epsilon\delta} = \{\nu \in \mathcal{D}_1 : \rho^L(G_{\delta_1}(\nu), G(\nu)) \leq \epsilon \text{ for } \delta_1 \leq \delta\}$; note we have suppressed x. Let $E_{\epsilon\delta;\beta} = \{\nu \in \mathcal{D}_1 : \rho_\phi(\nu, E_{\epsilon\delta}) \leq \beta\}$ be the β-neighborhood of $E_{\epsilon\delta}$. Let $\nu \in E_{\epsilon\delta;\beta}$ and $\delta_1 \leq \delta$. If $\nu' \in E_{\epsilon\delta}$, and $\rho_\phi(\nu', \nu) \leq 2\beta$, we have

$$\begin{aligned}
\rho^L(G_{\delta_1}(\nu), G(\nu)) &\leq \rho^L(G_{\delta_1}(\nu), G_{\delta_1}(\nu')) + \rho^L(G_{\delta_1}(\nu'), G(\nu')) + \rho^L(G(\nu'), G(\nu)) \\
&\leq 2\beta + \epsilon + 2\beta \\
&= \epsilon + 4\beta;
\end{aligned} \tag{8.9}$$

we have used the definition (7.21) of $\rho_\phi(\nu_1, \nu_2)$ in terms of $\rho^L(G_\delta(\nu_1), G_\delta(\nu_2))$. Thus

$$E_{\epsilon\delta;\beta} \subset E_{(\epsilon+4\beta)\delta}. \tag{8.10}$$

Given $\alpha \in \mathbf{N}$, let $\epsilon = \frac{1}{\alpha}$, $\delta = 2\sup_k \mu_{\alpha k}$, $\beta = 2\sup_k \operatorname{diam}(P_{\alpha k})$. By the Lemma ($\sim$ denotes complement)

$$\rho^L\Big(\sum_k \mu_{\alpha k} G(\nu_{\alpha k}), \sum_k \mu_{\alpha k} G_{\mu_{\alpha k}}(\nu_{\alpha k})\Big)$$

$$\leq \rho^L\Big(\sum_k \mu^*(P_{\alpha k}\cap E_{\epsilon\delta})G(\nu_{\alpha k}), \sum_k \mu^*(P_{\alpha k}\cap E_{\epsilon\delta})G_{\mu_{\alpha k}}(\nu_{\alpha k})\Big) \tag{8.11}$$

$$+\rho^L\Big(\sum_k \mu^*(P_{\alpha k}\cap \tilde{E}_{\epsilon\delta})G(\nu_{\alpha k}), \sum_k \mu^*(P_{\alpha k}\cap \tilde{E}_{\epsilon\delta})G_{\mu_{\alpha k}}(\nu_{\alpha k})\Big).$$

If $P_{\alpha k}\cap E_{\epsilon\delta} \neq \phi$, then $P_{\alpha k} \subset E_{(\epsilon+4\beta)\delta}$ by (8.10), since $\operatorname{diam}(P_{\alpha k}) \leq \frac{\beta}{2}$. Hence $\nu_{\alpha k} \in E_{(\epsilon+4\beta)\delta}$, and $\sup_{\{k:P_{\alpha k}\cap E_{\epsilon\delta}\neq\phi\}} \rho^L(G(\nu_{\alpha k}), G_{\mu_{\alpha k}}(\nu_{\alpha k})) \leq \epsilon + 4\beta \underset{\alpha\to\infty}{\to} 0$, so that the first term on the right of (8.11) vanishes as $\alpha \to \infty$, by Lemma 8.1.1. The second term, on the other hand, is bounded by $\mu^*(\tilde{E}_{\epsilon\delta})\sup_{\delta_1\leq 1;\nu\in\mathcal{D}_1} G_{\delta_1}(\nu,\infty)$. For $\nu_1, \nu_2 \in \mathcal{D}_1$,

$$G_\delta(\nu_1,\infty) - G_\delta(\nu_2,\infty) \leq \operatorname{diam}(\mathcal{D}_1) < \infty,$$

so that $\sup_{\delta_1\leq 1;\nu\in\mathcal{D}_1} G_{\delta_1}(\nu,\infty) < \infty$. Since $\mu^*(\tilde{E}_{\epsilon\delta}) \underset{\alpha\to\infty}{\to} 0$, this term vanishes as $\alpha\to\infty$, so that (8.8) vanishes as $\alpha \to \infty$. Together with (8.7) this proves (8.4b) (recall $\int_\Lambda G(X,x)d\mu(\lambda) = \int_{\mathcal{R}_{\text{ess}}(X)} G(\nu,x)d\mu^*(\nu)$).

We now prove (8.4a). By the mean value theorem,

$$\sum_k b_{\alpha k} + \int_{-\infty}^{\infty} \chi\, d\nu_{\alpha k}\Big(\frac{x+b_{\alpha k}}{\phi_{\alpha k}}\Big)$$

$$= \sum_k \int_{-\infty}^{\infty} \chi\, d\nu_{\alpha k}\Big(\frac{x}{\phi_{\alpha k}}\Big) + \sum_k b_{\alpha k}\Big(\int_{-\infty}^{\infty}(1-\chi'(x+b^*_{\alpha k}))d\nu_{\alpha k}\Big(\frac{x}{\phi_{\alpha k}}\Big)\Big),$$

for some $|b^*_\alpha| < |b_\alpha|$. By the arguments for (8.6), the last term vanishes as $\alpha\to\infty$, and we are left with proving

$$\sum_k \int_{-\infty}^{\infty} \chi\, d\nu_{\alpha k}\Big(\frac{x}{\phi_{\alpha k}}\Big) \to \gamma;$$

which also follows along the same lines as the first part of the proof.

We now prove the inverse of the above, namely, that (8.4) is necessary for convergence. Suppose first that for some $\nu \in \operatorname{supp}\mu^*$, $G(\nu)$ fails to exist. Define $\epsilon > 0$ by

$$\frac{\epsilon}{\mu^*(\mathcal{D}_1)} = \lim_{\delta\to 0}\ \sup_{\delta_1,\delta_2<\delta} \rho^L(G_{\delta_1}(\nu), G_{\delta_2}(\nu)). \tag{8.12}$$

Let $N = \{\nu' \in \mathcal{D}_1 : \rho_\phi(\nu, \nu') \leq \frac{\epsilon}{4}\}$. We show that ${}_L\int_N \nu\phi(d\mu^*)$ fails to exist, and hence (see Proposition 7.6) the integral over $\mathcal{D}_1$ fails to exist. To this end, we may assume that $N = \mathcal{D}_1$. Let P_α be an infinitesimal sequence of partitions of $\mathcal{D}_1$. In the following, we may assume the partitions are even ($\mu_{\alpha i} = \mu_{\alpha j}$) without loss, by the arguments of Theorem 7.11 (i.e., the total measure of those elements $P_{\alpha i}$ for which $\mu^*(P_{\alpha i})$ does not equal the common value can be made arbitrarily small). Then (letting δ_α be the common value of $\mu_{\alpha k}$)

$$\rho^L\Big(\sum_k \mu_{\alpha k} G_{\mu_{\alpha k}}(\nu_{\alpha k}), \mu^*(\mathcal{D}_1) G_{\delta_\alpha}(\nu)\Big) = \rho^L\Big(\sum_k \delta_\alpha G_{\delta_\alpha}(\nu_{\alpha k}), \sum_k \delta_\alpha G_{\delta_\alpha}(\nu)\Big)$$

$$\leq \max\Big(\sum_k \delta_\alpha \rho^L(G_{\delta_\alpha}(\nu_{\alpha k}), G_{\delta_\alpha}(\nu)), \sup_k \rho^L(G_{\delta_\alpha}(\nu_{\alpha k}), G_{\delta_\alpha}(\nu))\Big) \tag{8.13}$$

$$\leq \max\Big(\frac{\epsilon}{4}\mu^*(\mathcal{D}_1), \frac{\epsilon}{4}\Big) = \frac{\epsilon}{4},$$

assuming (without loss) that $\mu^*(\mathcal{D}_1) \leq 1$.

Again without loss of generality, we may assume by (8.12) that the sequence δ_α is such that

$$\mu^*(\mathcal{D}_1) \lim_{\alpha\to\infty} \sup_{\alpha_1,\alpha_2 \geq \alpha} \rho^L(G_{\delta_{\alpha_1}}(\nu), G_{\delta_{\alpha_2}}(\nu)) = \epsilon \tag{8.14}$$

Hence by (8.13),

$$\lim_{\alpha\to\infty} \sup_{\alpha_1,\alpha_2 \geq \alpha} \rho^L\Big(\sum_k \mu_{\alpha_1 k} G_{\mu_{\alpha_1 k}}(\nu_{\alpha_1 k}), \sum_k \mu_{\alpha_2 k} G_{\mu_{\alpha_2 k}}(\nu_{\alpha_2 k})\Big) \geq \frac{\epsilon}{2}, \tag{8.15}$$

so that $\sum_k \mu_{\alpha k} G_{\mu_{\alpha k}}(\nu_{\alpha k})$ fails to converge, and by (8.6), so does $\sum_k \mu_{\alpha k} G_{\mu_{\alpha k}}(\nu_{\alpha k}(x + b_{\alpha k}))$. (The last term in (8.6) still vanishes as $\alpha \to \infty$). Thus ${}_L\int_{\mathcal{D}_1} \nu\phi(d\mu^*)$ fails to exist by Corollary 7.3.1.

Now assume γ_ν fails to exist. If the weak limit of $\theta d\nu_\delta$ exists, then

$$\int_{-\infty}^{\infty} \chi^*(x)\, d\nu_\delta(x) = \int_{-\infty}^{\infty} \chi(x)\, d\nu_\delta(x) + \int_{-\infty}^{\infty} f(x)\theta(x)\, d\nu_\delta(x), \tag{8.16}$$

where $f \in C_B(\mathbf{R})$, so that the left side of (8.16) also fails to converge as $\delta \to 0$. From here on the argument using Corollary 7.3.1 is the same as above, and again ${}_L\int \nu\phi(d\mu^*)$ fails to converge. This completes the proof. ∎

The above conditions can be simplified to a large extent.

DEFINITIONS 8.2: For $\eta' \in \mathbf{R}^+$, $\mathcal{A}_{\eta'}$ denotes those functions $\phi(\delta)$ defined for δ small and positive satisfying $\frac{\phi(\delta)}{\delta^{\eta'}} \underset{\delta\to 0}{\to} C \in \mathbf{R}$. Let $\mathcal{A} = \bigcup_{\eta' \geq \frac{1}{2}} \mathcal{A}_{\eta'}$. Define $\mathbf{R}^{\pm} \equiv \mathbf{R} \sim \{0\}$.

THEOREM 8.3: *Let X, ϕ, and μ satisfy the hypotheses of Theorem 8.1. Then $Y = {}_L\int_\eta X\phi(d\mu)$ exists if and only if*

(i) $\phi \in \mathcal{A}_{\eta'}$ for some $\eta' \geq \frac{1}{2}$

(ii) for ν in the essential range of X, ν_δ converges weakly to a measure on $\mathbf{R}^\pm$ as $\delta \to \infty$.

(iii) for $\nu \in \operatorname{supp}\mu^*$, (a) if $\eta' = \frac{1}{2}$ then $\mathcal{V}(\nu) < \infty$; (b) if $\frac{1}{2} \leq \eta' < 1$, $\mathcal{E}(\nu) = 0$; and (c) if $\eta' = 1$, $\lim_{\eta\to\infty} \int_{-M}^{M} x\,d\nu$ exists.

Note that the weak convergence condition above means ν_δ converges to (possibly infinite) measures on $\mathbf{R}^+$ and $\mathbf{R}^-$, individually.

Proof: We first show the above imply the necessary and sufficient conditions of the previous theorem. Let $Q\nu \equiv \text{w-}\lim_{\delta\to 0}\nu_\delta \equiv \text{w-}\lim \frac{1}{\delta}\nu\left(\frac{x}{\delta^{\eta'}}\right)$. Simple scaling arguments (replace x by cx and let $\delta \to 0$) show that $Q\nu$ is homogeneous, and that

$$d(Q\nu) = \left(c_1|x|^{-\eta-1} + c_2|x|^{-\eta-1}\frac{x}{|x|}\right)dx \qquad (x \in \mathbf{R}^\pm), \tag{8.17}$$

where $\eta = \frac{1}{\eta'}$. For $x > 0$, let $F^+(x) = \nu[x,\infty)$; $F^-(-x) = \nu(-\infty,-x)$. Then since

$$\frac{1}{\delta}F^+\left(\frac{x}{\delta^{\eta'}}\right) \underset{\delta\to 0}{\to} \frac{c_1+c_2}{\eta}x^{-\eta},$$

it follows that

$$x^\eta F^+(x) \underset{x\to\infty}{\to} \frac{c_1+c_2}{\eta},$$

and similarly

$$(x)^\eta F^-(-x) \underset{x\to\infty}{\to} \frac{c_1-c_2}{\eta}. \tag{8.18}$$

If $\eta' > \frac{1}{2}$, then for $\nu \in \operatorname{supp}\mu^*$,

$$\int_{-1}^{1} x^2 d\nu_\delta = M^{\eta-2}\int_{-M}^{M} x^2 d\nu(x), \tag{8.19}$$

where $M = \frac{1}{\delta}$; (8.19) is easily shown to converge using the asymptotics of F^+ and F^-, after integration by parts. Hence $x^2 d\nu_\delta$ converges weakly on $[-1,1]$ and thus $\theta(x)d\nu_\delta$ does on $\mathbf{R}$, so that the first weak limit in (8.3) exists. To examine the same limit when $\eta' = \frac{1}{2}$, notice that (8.19) now converges by the finiteness of $\mathcal{V}(\nu)$.

We prove existence of the second limit in (8.3). If $\eta' \geq 1$, the limit follows from the convergence of

$$\int_{-1}^{1} x\,d\nu_\delta = M^{\eta-1}\int_{-M}^{M} x\,d\nu(x), \tag{8.20}$$

via integration by parts as before. When $\frac{1}{2} \leq \eta' < 1$,

$$M^{\eta-1}\int_{-M}^{M} x\,d\nu = -M^{\eta-1}\left[\int_{M+}^{\infty} x\,d\nu + \int_{-\infty}^{-M^-} x\,d\nu\right],$$

and the latter again converges via integrations by parts. This completes the proof of sufficiency.

We now assume (8.3) to hold, and prove $(ii)-(iii)$ ((i) is clearly necessary by Theorem 7.11). First, (ii) follows immediately from convergence of $G(\nu;x)$, since weak convergence of $\theta d\nu_\delta$ on $\mathbf{R}^\pm$ implies the same for $d\nu_\delta$. To prove (iii), first let $\eta' = \frac{1}{2}$. Since $\theta d\nu_\delta$ converges weakly by hypothesis,

$$\lim_{\delta\to 0}\int_{-1}^{1} x^2 d\nu_\delta = \lim_{M\to\infty}\int_{-M}^{M} x^2 d\nu$$

converges, proving (iii), (a). Similarly, if $\frac{1}{2} \leq \eta' < 1$, then

$$\lim_{\delta\to 0}\int_{-1}^{1} x\,d\nu_\delta = \lim_{M\to\infty} M^{\eta-1}\int_{-M}^{M} x\,d\nu. \tag{8.21}$$

Since this converges (by 8.3) and $\eta - 1 > 0$, it follows that $\lim_{M\to\infty}\int_{-M}^{M} x\,d\nu = 0$. To show that ν indeed has a first moment, we show $\int_0^\infty x\,d\nu < \infty$ (since the argument on $\mathbf{R}^-$ is the same). Let G be a monotone function, $G(0) = 0$, defined by $dG = x\,d\nu$. By (8.19) and (8.3), we have convergence as $M \to \infty$ of

$$M^{\eta-2}\int_0^M x^2 d\nu = M^{\eta-1}(G(M) - A(M)),$$

where

$$A(M) = \frac{1}{M}\int_0^M G(x)\,dx. \tag{8.22}$$

Hence

$$G(M) - A(M) \leq h(M) \qquad (M > 0),$$

where $h(M)$ is a smooth positive function, with $h(0) = 0$, $h'(0) \leq C_1$, $h(M) = C_2 M^{1-\eta}$ for $M \geq 1$; note that $G(x) \leq x$. There exists a monotone function G^*, $G^*(0) = 0$, satisfying

$$G^*(M) - A^*(M) = h(M),$$

where G^* and A^* are related by (8.22). The equation for $M \geq 1$,

$$G^*(M) - A^*(M) = C_2 M^{1-\eta} \tag{8.23}$$

is solved by differentiating once, giving

$$G^*(M) = C_3 + C_2\left(\frac{2-\eta}{1-\eta}\right)M^{-\eta} \qquad (M \geq 1),$$

so that G^* is bounded. On the other hand,

$$(G^* - G) - (A^* - A) \geq 0 \tag{8.24}$$

on $\mathbf{R}^+$; this implies $(G^* - G)(M) \geq (G^* - G)(0)$ for $M \geq 0$, so G is bounded, proving that ν has a finite first moment on $\mathbf{R}^+$. The proof is of course the same on $\mathbf{R}^-$. Thus necessity of (*iii*) (b) is proved; necessity of (*iii*) (c) for convergence of the second integral in (8.3) is clear, and this completes the proof. ∎

We now consider the general (non-finite) Lebesgue integral on a measure space $(\Lambda, \mathcal{B}, \mu)$. Recall (8.2).

LEMMA 8.4.1: *Let X_i be a sequence of independent r.v.'s, and $b_i = \mathcal{E}\chi^*(X_i)$. The sum $\sum_{i=1}^{\infty} X_i$ converges order independently if and only if $\sum |b_i|$ and $\sum \mathcal{E}\theta(X_i - b_i)$ converge.*

Proof: Assume the numerical series converge absolutely, and write, using the mean value theorem,

$$\sum_i \mathcal{E}\theta(X_i + b_i) = \sum_i \mathcal{E}\theta(X_i) + \sum_i b_i \mathcal{E}\theta'(X_i + b_i^*), \tag{8.25}$$

where $|b_i^*| \leq |b_i|$ with probability one. It is easy to see the X_i are infinitesimal, and in particular that $\mathcal{E}\theta'(X_i+b_i) \underset{i\to\infty}{\to} 0$, so by (8.25) $\sum \mathcal{E}\theta(X_i)$ converges. From this it follows (with convergence of $\sum |b_i|$) that $\sum \mathcal{E}\chi(X_i)$ converges as well, and sufficiency follows by Lemma 7.3.1. Conversely, using Lemma 7.3.1, if $\sum \mathcal{E}\theta(X_i)$ and $\sum \mathcal{E}\chi^*(X_i)$ converge absolutely, so does $\sum \mathcal{E}\chi(X_i)$, and $\sum \mathcal{E}\theta(X_i - b_i)$ converges by (8.25). ∎

Recall that, for a measure ν on $\mathbf{R}$ and $\phi : \mathbf{R}^+ \to \mathbf{R}^+$, the measures ν_δ and $G\nu$ are

$$\nu_\delta = \frac{1}{\delta}\nu\left(\frac{x}{\phi(\delta)}\right), \qquad d(G\nu) = \text{w-}\lim_{\delta\to 0} \theta d\nu_\delta, \tag{8.26}$$

and

$$\gamma_\nu = \lim_{\delta\to 0}\int_{-\infty}^{\infty} \chi(x)d\nu_\delta(x), \qquad \gamma_\nu^* = \lim_{\delta\to 0}\int_{-\infty}^{\infty} \chi^*(x)d\nu_\delta(x) \tag{8.27}$$

(see (8.2)), when the limits exist. If X is an r.v., GX and γ_X are defined analogously by the distribution ν of X. If μ is a measure, $|\mu|$ denotes total measure.

THEOREM 8.4: *Given $\phi : \mathbf{R}^+ \to \mathbf{R}^+$, and $X : \Lambda \to \mathcal{D}$, a random variable-valued function on the measure space Λ, the integral $Y \equiv {}_L\int_\Lambda X(\lambda)\phi(d\mu(\lambda))$ exists if and only if (1) $G\nu$ and γ_ν exist for every ν in the essential range of X, and (2) $\int_\Lambda |GX(\lambda)| d\mu(\lambda)$ and $\int_\Lambda \left|\gamma_{X(\lambda)}\right| d\mu(\lambda)$ converge. Furthermore, the pair $\left(\int_\Lambda \gamma_X d\mu, \int_\Lambda GX\, d\mu\right)$ is the Lévy-Khinchin transform of Y.*

Note that we make no restriction on $\mathcal{D}$, i.e., X may be *any* measurable map into the space of probability distributions. This theorem, together with Proposition 7.6, shows that an integrable X induces a countably additive measure-valued measure $\mathcal{M}$ defined by $\mathcal{M}(B) = \int_B X\phi(d\mu)$.

Proof: Assume (1) and (2) hold. Let μ^* be the measure induced on $\mathcal{D}$ by X. Since $|\gamma^*_\nu| \leq c_1|\gamma_\nu| + c_2|G\nu|$ for some $c_1, c_2 > 0$, $\int_{\mathcal{D}} \left|\gamma^*_{X(\lambda)}\right| d\mu^*$ converges as well. Let $\{P_i\}_{i\in I}$ be an at most countable partition of $\mathcal{D}_1 = \mathcal{R}_{\mathrm{ess}}(X)$ into sets of finite measure and diameter. Let

$$\mathcal{D}_{11} = \{\nu \in \mathcal{D}_1 : \gamma^*_\nu \geq 0\}, \qquad \mathcal{D}_{12} = \{\nu \in \mathcal{D}_1 : \gamma^*_\nu < 0\}.$$

Assume without loss that P respects the partition $\mathcal{D}_{11}, \mathcal{D}_{12}$ of $\mathcal{D}_1$. Let $\{P_{ij}\}_j$ be an at most countable partition of P_i whose elements have finite measure and diameter, with $\nu_{ij} \in P_{ij}$, $\mu_{ij} = \mu^*(P_{ij})$, $\phi_{ij} = \phi(\mu_{ij})$. Let $b_{ij} = \int_{-\infty}^{\infty} \chi^*(x) d\nu_{ij}\left(\frac{x}{\phi_{ij}}\right)$. By the proof of Theorem 8.1, $\sum_j |b_{ij}|$ can be made to differ from $\int_{P_i} |\gamma^*_\nu| d\mu^*(\nu)$ by arbitrarily little; hence the sub-partitions P_{ij} can be chosen so that $\sum_{i\in I,j} |b_{ij}|$ converges. Given $i \in I$, for a sufficiently small sub-partition of $\{P_{ij}\}_j$ (still denoted by $\{P_{ij}\}$; P_i is unchanged), $\sum^*_j \nu_{ij}\phi_{ij}$ can be made arbitrarily close to ${}_L\int_{P_i} \nu\phi(d\mu^*(\nu))$. Hence by Theorem 8.1,

$$\sum_j \int_{-\infty}^{\infty} \theta\, d\nu_{ij}\left(\frac{x + b_{ij}}{\phi_{ij}}\right) - \int_{P_i} |G\nu| d\mu^* \tag{8.28}$$

can be made arbitrarily small. Therefore by condition (2), there exists a subpartition $\{P_{ij}\}_{i,j}$ such that $\sum_{i,j} b_{ij}$ and $\sum_{i,j} \int_{-\infty}^{\infty} \theta d\nu_{ij}\left(\frac{x+b_{ij}}{\phi_{ij}}\right)$ converge absolutely under P_{ij}, as well as any finer partition. Thus, by the Lemma, $\sum^*_i \left(\sum^*_j \nu_{ij}\phi_{ij}\right)$ converges order independently. Since the inner sum approximates ${}_L\int_{P_i} \nu\phi(d\mu^*)$ arbitrarily well, the sum $\sum^*_i {}_L\int_{P_i} \nu\phi(d\mu^*) = {}_L\int_\Lambda X(\lambda)\phi(d\mu(\lambda))$ is also order independent. Since any two partitions have a minimal common refinement, this sum is also independent of $\{P_i\}$.

Conversely, if ${}_L\int_\Lambda X\phi(d\mu)$ exists, we claim $\int_\Lambda |GX| d\mu < \infty$. For if the latter is false, let $\{P_i\}_{i\in I}$ be an at most countable partition of Λ into sets of finite measure and diameter. The sum $\displaystyle\sum_i \int_{P_i} |GX| d\mu$ diverges; and therefore $\displaystyle\sum_i \int_{P_i} X\phi(d\mu)$ must also fail to converge by the linearity and continuity of the Lévy-Khinchin transform (Remarks after Theorem 7.1). Thus the integral fails to exist. A similar contradiction obtains if $\int_\Lambda |\gamma_\nu| d\mu^* = \infty$.

Finally, if $Y = {}_L\int_\Lambda X\phi(d\mu)$ exists, then $Y = \sum_i Y_i$, where

$$Y_i = {}_L\int_{P_i} X\phi(d\mu)$$

and P_i is a sequence of subsets of Λ with finite measure and diameter. Thus, the Lévy-Khinchin transform of Y is the sum of those of Y_i. This together with the final assertion in Theorem 8.1 completes the proof.∎

We remark that conditions (1) can naturally be replaced by conditions (i)–(iii) of Theorem 8.3.

The following lemma is proved in Loève [Lo, §23].

LEMMA 8.5.1: *Let $\alpha_n \in \mathbf{R}$, and Ψ_n be finite Borel measures on $\mathbf{R}$. If $it\alpha_n + \int_{-\infty}^{\infty}\left(e^{itx} - 1 - \frac{itx}{1+x^2}\right)\frac{(1+x^2)}{x^2}d\Psi_n$ converges to a function continuous at the origin, then $\{\alpha_n\}_n$ converges and $\{\Psi_n\}_n$ converges weakly.*

THEOREM 8.5: *The integral $Y = {}_L\int_\Lambda X\phi(d\mu)$ exists if and only if*
(i) $\psi_\nu(t) = \lim_{\delta\to 0}\frac{1}{\delta}(\Phi_\nu(\phi(\delta)t) - 1)$ exists and is continuous at $t = 0$ for $\nu \in \mathcal{R}_{\text{ess}}(X)$, where Φ_ν is the characteristic function of ν
(ii) the integral

$$\psi(t) = \int_\Lambda \psi_{X(\lambda)}(t)d\mu(\lambda) \tag{8.29}$$

converges absolutely.
In this case Y has characteristic function e^ψ.

Proof: If Y exists, then according to Theorem 8.4, so do γ_ν and $G\nu$ for $\nu \in \mathcal{R}_{\text{ess}}(X)$. Furthermore, if $\nu \in \mathcal{R}_{\text{ess}}(X)$,

$$\lim_{\delta\to 0}\int(e^{itx} - 1)d\nu_\delta(x) = i\gamma_\nu t + \int\left(e^{ixt} - 1 - \frac{ixt}{1+x^2}\right)\frac{1+x^2}{x^2}d(G\nu)(x)$$

is clearly continuous at 0, proving (i). According to (7.1), Y has the log ch.f.

$$\psi_1(t) = i\gamma t + \int_{-\infty}^{\infty}\left(e^{ixt} - 1 - \frac{ixt}{1+x^2}\right)\left(\frac{1+x^2}{x^2}\right)dG \tag{8.30}$$

where, by Theorem 8.4,

$$\gamma = \int_D \gamma_\nu d\mu^*, \qquad G = \int_D G\nu\, d\mu^*,$$

and μ^* is the induced measure on $\mathcal{D}$. Since the integrand in (8.30) is bounded, we can interchange integrals, so

$$\psi_1(t) = \int_{\mathcal{D}} \left[i\gamma_\nu t + \int_{-\infty}^{\infty} \left(e^{ixt} - 1 - \frac{ixt}{1+x^2} \right) \left(\frac{1+x^2}{x^2} \right) d(G\nu)(x) \right] d\mu^*(\nu)$$

$$= \int_{\mathcal{D}} \lim_{\delta \to 0} \left[it \int_{-\infty}^{\infty} \frac{x}{1+x^2} d\nu_\delta + \int_{-\infty}^{\infty} \left(e^{ixt} - 1 - \frac{ixt}{1+x^2} \right) d\nu_\delta \right] d\mu^*(\nu) \tag{8.31}$$

$$= \int_{\mathcal{D}} \left[\lim_{\delta \to 0} \frac{1}{\delta} \int_{-\infty}^{\infty} (e^{ix\phi t} - 1) d\nu(x) \right] d\mu^*(\nu) = \psi(t).$$

It is clear that the integrals converge absolutely.

Conversely, assume that (i) and (ii) hold. If $\nu \in \mathcal{R}_{\text{ess}}(X)$,

$$\lim_{\delta \to 0} \int (e^{ixt} - 1) d\nu_\delta = \lim_{\delta \to 0} it \int \chi \, d\nu_\delta + \int \left(e^{itx} - 1 - \frac{itx}{1+x^2} \right) \left(\frac{1+x^2}{x^2} \right) \theta(x) d\nu_\delta$$

is continuous at 0, so that by the Lemma, $\gamma_\nu = \lim \int \chi d\nu_\delta$ and $dG\nu = \text{w-}\lim \theta d\nu_\delta$ exist. Let $\{P_i\}_{i \in I}$ be a countable partition of Λ into sets of finite measure, on each on which X is essentially bounded. This can be done by appropriately partitioning $\mathcal{R}_{\text{ess}}(X)$. Let $\Lambda_n = \bigcup_{i=1}^{n} P_i$. By Theorem 8.4, $\int_{\Lambda_n} X \phi(d\mu)$ exists for all n, and by part one of this proof, if ψ_n is its log ch.f., then

$$\psi_n = \int_{\Lambda_n} \psi_{X(\lambda)} d\mu(\lambda). \tag{8.32}$$

Clearly, $\int_{\Lambda_n} X \phi(d\mu)$ converges weakly as $n \to \infty$ to an r.v. Y with ch.f. $\psi(t)$. Thus Y is independent of the choice of $\{P_i\}_{i \in I}$, and is the desired integral. ∎

§8.2 The Riemann Integral

Let (Λ, σ) be a metric space with σ-finite Borel measure μ, and $X\colon \Lambda \to \mathcal{D}$ be an r.v.-valued function. The Riemann integral of X is defined with respect to partitions of Λ rather than $\mathcal{D}$. We might again try to duplicate the elegance of the Lebesgue theory in the Riemann integral, but we approach the latter with a view to utility, namely, to physically motivated applications. With this intent, we define the Riemann integral using global partitions of Λ, rather than patching integrals over sets of finite measure whose range under X is bounded. The former definition will dovetail with that of the R^*-integral.

DEFINITION 8.6: Let $\{P_\alpha\}_{\alpha=1}^{\infty}$ be an infinitesimal sequence of partitions of Λ. Select a function $\phi : \mathbf{R}^+ \to \mathbf{R}^+$. Let $\lambda_{\alpha i} \in P_{\alpha i}$, and assume

$$\sum_i X(\lambda_{\alpha i}) \phi(\mu_{\alpha i}) \underset{\alpha \to \infty}{\Rightarrow} Y,$$

where $\mu_{\alpha i} = \mu(P_{\alpha i})$, and the $\{X(\lambda)\}_{\lambda \in \Lambda}$ are independent. If Y is independent of $\{P_\alpha\}$, then Y is the (ϕ-) *Riemann integral* of X (with respect to the non-linear measure $\phi(d\mu)$),

$$Y = {}_R\int_\Lambda X(\lambda)\, \phi(d\mu(\lambda)) \tag{8.33}$$

THEOREM 8.7: *If $X(\lambda)$ is ϕ-Riemann integrable then it is ϕ-Lebesgue integrable, and the integrals coincide.*

Proof: Recall (7.21). We begin by assuming μ is finite and X is ρ_ϕ-bounded. Let P_α be an infinitesimal net of partitions of Λ. Recall the Lebesgue integral may be equivalently defined by allowing partitions of $\mathcal{D}$ to subdivide individual elements, even ones of measure 0. Under this more general (but equivalent) definition, there exists a partition sequence P^*_α of $\mathcal{D}$ such that $X(P_{\alpha i}) = P^*_{\alpha i}$. As before, we may use the arguments of Theorem 7.11 to assume (without subsequent loss) that the partitions P_α (P^*_α) are even with respect to μ ($\mu^* = \mu X^{-1}$), i.e., $\mu(P_{\alpha_i}) = \mu(P_{\alpha_j})$; again the measure of those elements with odd (unequal) measure can be made arbitrarily small by taking sufficiently small sub-partitions. By taking sub-partitions, we may also assume that $P^*_{\alpha i}$ vanish in diameter uniformly as $\alpha \to \infty$.

We proceed by contraposition, assuming the Lebesgue integral fails to exist. In this case it suffices to show the Riemann integral $\int_{\Lambda_1} X\phi(d\mu)$ fails to exist for any $\Lambda_1 \subset \Lambda$ with positive measure, by arguments used in proving Proposition 7.6. By our assumption and Theorem 8.1, either $G\nu$ or γ_ν fails to exist for some $\nu \in \mathcal{R}_{\text{ess}}$. Assume $G\nu$ does not exist. At this point, using the above net P^*_α of partitions, the argument becomes exactly analogous to that after (8.12), so we omit the details. The same argument works if γ_ν fails to exist, yielding the result when μ is finite and X is ρ_ϕ–bounded. The fact that the two integrals coincide in this case follows from correspondence of the Riemann sums over the partitions P_α and P^*_α.

If $\mu(\Lambda)$ is infinite or X is unbounded, and ${}_R\int_\Lambda X\phi(du)$ exists, then the integral also exists over any subset $\Lambda_i \subset \Lambda$ with positive measure. Since the essential range of X is σ-finite, it is separable. Let $\{\mathcal{D}_i\}$ be an at most countable partition of $\mathcal{D}$, with $\mu^*(\mathcal{D}_i)$ and $\text{diam}(\mathcal{D}_i) < \infty$. If $\Lambda_i = X^{-1}(\mathcal{D}_i)$, then ${}_L\int_{\Lambda_i} X\phi(d\mu)$ exists, and we must show that $\sum_i {}_L\int_{\Lambda_i} X\phi(d\mu)$ converges order independently to ${}_R\int_\Lambda X(d\mu)$. This follows from the fact that ${}_L\int_{\Lambda_i} X\phi(d\mu)$ can be approximated arbitrarily well by Riemann sums over partitions of Λ_i whose total sum over i approximates ${}_R\int_\Lambda X\phi(d\mu)$. ∎

Recall that for $\nu \in \mathcal{D}$,

$$G_\delta(\nu; x) = \frac{1}{\delta}\int_{-\infty}^{x} \theta(y)\, d\nu\!\left(\frac{y}{\phi(\delta)}\right); \qquad \gamma_\delta(\nu) = \frac{1}{\delta}\int_{-\infty}^{\infty} \chi(y)\, d\nu\!\left(\frac{y}{\phi(\delta)}\right).$$

We now prove

THEOREM 8.8: *Suppose*
(i) $X : \Lambda \to \mathcal{D}$ *is Lebesgue integrable and* ρ_ϕ*–continuous,*
(ii) the function $g_{\delta,\epsilon}(\lambda) \equiv \sup_{\substack{\delta_1 \le \delta \\ \sigma(\lambda_1,\lambda)\le\epsilon}} G_{\delta_1}(X(\lambda_1),\infty) + |\gamma_{\delta_1}(X(\lambda_1))|$ *is in* $L^1(\mu(\lambda))$ *for some* $\epsilon, \delta > 0$.
Then X *is Riemann integrable.*

Proof: We prove this for partitions which do not subdivide points (the proof therefore does not directly work for measure spaces with atoms). The general case (which allows partition elements which overlap on atoms) follows with small modifications.

Let P_α be an infinitesimal net of partitions of Λ, and $\lambda_{\alpha k} \in P_{\alpha k}$. Let $X_{\alpha k} = X(\lambda_{\alpha k})$, $\mu_{\alpha k} = \mu(P_{\alpha k})$, $\phi_{\alpha k} = \phi(\mu_{\alpha k})$. We first verify that the double sequence $\{\phi_{\alpha k} X_{\alpha k}\}$ is infinitesimal. Suppose it is not. Taking a subsequence if necessary, we may assume without loss of generality that

$$\mu_{\alpha k_\alpha} G_{\mu_{\alpha k_\alpha}}(X_{\alpha k_\alpha}, \infty) \ge \epsilon_1$$

for some $\epsilon_1 > 0$, and some choice of k_α for each α. Therefore if $\delta, \epsilon > 0$,

$$g_{\delta,\epsilon}(\lambda) \ge \frac{\epsilon_1}{\mu_{\alpha k_\alpha}} \qquad (\sigma(\lambda, \lambda_{\alpha k_\alpha}) \le \epsilon) \tag{8.34}$$

for α sufficiently large that $\mu_{\alpha k_\alpha} \le \delta$. Furthermore, again taking subsequences if necessary, we assume that $\mu_{\alpha k_\alpha} \le \frac{1}{2}\mu_{(\alpha-1)k_{\alpha-1}}$. Then if $B_\alpha = B_\epsilon(\lambda_{\alpha k_\alpha})$ (the ϵ-ball about $\lambda_{\alpha k_\alpha}$),

$$\begin{aligned}
\int_\Lambda g_{\delta,\epsilon}\, d\mu(\lambda) &\ge \left(\int_{B_1} + \int_{B_2 \sim B_1} + \int_{B_3 \sim (B_1 \cup B_2)} + \ldots\right) g_{\delta,\epsilon}\, d\mu \\
&\ge \frac{\epsilon_1}{\mu_{1k_1}}\mu(B_1) + \left(\frac{\epsilon_1}{\mu_{2k_2}} - \frac{\epsilon_1}{\mu_{1k_1}}\right)\mu(B_2) + \left(\frac{\epsilon_1}{\mu_{3k_3}} - \frac{\epsilon_1}{\mu_{2k_2}}\right)\mu(B_3) + \cdots \\
&\ge \frac{\epsilon_1}{\mu_{1k_1}}\mu(B_1) + \frac{1}{2}\frac{\epsilon_1}{\mu_{2k_2}}\mu(B_2) + \frac{1}{2}\frac{\epsilon_1}{\mu_{3k_3}}\mu(B_3) + \cdots \\
&= \infty,
\end{aligned}$$

since $\mu(B_\alpha) \ge \mu_{\alpha k_\alpha}$ for α sufficiently large; this gives the desired contradiction. Thus, $\{\phi_{\alpha k} X_{\alpha k}\}$ is infinitesimal.

To prove the theorem, according to Corollary 7.3.1, it suffices to prove the convergence

$$\sum_k \int_{-\infty}^{x} \theta\, d\nu_{\alpha k}\left(\frac{y + b_{\alpha k}}{\phi_{\alpha k}}\right) \underset{\alpha\to\infty}{\Rightarrow} G(x); \tag{8.35a}$$

$$\sum_k b_{\alpha k} \underset{\alpha\to\infty}{\to} \gamma', \tag{8.35b}$$

for some $G:\mathbf{R}\to\mathbf{R}$ a multiple of a d.f., and some $\gamma'\in\mathbf{R}$, both independent of $\{P_\alpha\}$. Here, $\nu_{\alpha k}$ is the distribution of $X_{\alpha k}$, and

$$b_{\alpha k}=\int_{-\infty}^{\infty}\chi^* d\nu_{\alpha k}\left(\frac{x}{\phi_{\alpha k}}\right), \tag{8.36}$$

with χ^* given by (7.9). We have

$$\sum_k\int_{-\infty}^{x}\theta\, d\nu_{\alpha k}\left(\frac{y+b_{\alpha k}}{\phi_{\alpha k}}\right)=\int_\Lambda H_\alpha(\lambda;x)\,d\mu(\lambda),$$

where

$$H_\alpha(\lambda;x)=\frac{1}{\mu_{\alpha k}}\int_{-\infty}^{x}\theta d\nu_{\alpha k}\left(\frac{y+b_{\alpha k}}{\phi_{\alpha k}}\right)\qquad(\lambda\in P_{\alpha k}).$$

Since X is ρ_ϕ-continuous, if $\lambda\in\text{supp}(\mu)$, then $X(\lambda)\in\mathcal{R}_{\text{ess}}(X)$, so that by (1) of Theorem 8.4,

$$H_\alpha(\lambda;x)\underset{\alpha\to\infty}{\Rightarrow}H(\lambda;x), \tag{8.37}$$

where $H(\lambda;x)$ is a multiple of a d.f. in x for each λ. Furthermore,

$$\begin{aligned}H_\alpha(\lambda;x)&\le H_\alpha(\lambda;\infty)\\&=\frac{1}{\mu_{\alpha k}}\int_{-\infty}^{\infty}\theta\,d\nu_{\alpha k}\left(\frac{y}{\phi_{\alpha k}}\right)+\frac{b_{\alpha k}{}^*}{\mu_{\alpha k}}\int_{-\infty}^{\infty}\theta'(x)\,d\nu_{\alpha k}\left(\frac{y}{\phi_{\alpha k}}\right)\quad(\lambda\in P_{\alpha k}),\end{aligned} \tag{8.38}$$

where $|b^*_{\alpha k}|\le|b_{\alpha k}|$ is determined by the mean value theorem. The first term on the right is clearly dominated by $g_{\delta,\epsilon}(\lambda)\in L^1(\mu)$ for α sufficiently large. Since the double sequence $\nu_{\alpha k}\left(\frac{x}{\phi_{\alpha k}}\right)$ is infinitesimal and $\theta'(0)=0$, the second term on the right is eventually dominated by

$$H^{(2)}_\alpha(\lambda;\infty)=\frac{b_{\alpha k}}{\mu_{\alpha k}}=\frac{1}{\mu_{\alpha k}}\int_{-\infty}^{\infty}\chi^* d\nu_{\alpha k}\left(\frac{x}{\phi_{\alpha k}}\right)\le Kg_{\epsilon,\delta}(\lambda)\qquad(\lambda\in P_{\alpha k}), \tag{8.39}$$

for α sufficiently large, with K independent of α. Thus, for α sufficiently large,

$$H_\alpha(\lambda;u)\le(1+K)g_{\delta,\epsilon}(\lambda)\in L^1. \tag{8.40}$$

By (8.37), if $h(x)\in C_c^\infty(\mathbf{R})$,

$$\int_{-\infty}^{\infty}H_\alpha(\lambda,x)h(x)dx\underset{\alpha\to\infty}{\to}\int_{-\infty}^{\infty}H(\lambda,x)h(x)dx,$$

so that by (8.40), the dominated convergence therorem, and the Fubini theorem,

$$\int_{-\infty}^{\infty}\int_\Lambda H_\alpha(\lambda;x)\,d\mu(\lambda)\phi(x)dx\underset{\alpha\to\infty}{\to}\int_{-\infty}^{\infty}\int_\Lambda H(\lambda;x)\,d\mu(\lambda)\phi(x)dx. \tag{8.41}$$

Since $h \in C_c^\infty(\mathbf{R})$ is arbitrary, it follows that

$$\sum_k \int_{-\infty}^{x} \theta \, d\nu_{\alpha k}\left(\frac{y + b_{\alpha k}}{\phi_{\alpha k}}\right) = \int_\Lambda H_\alpha(\lambda; x) \, d\mu(\lambda) \underset{\alpha\to\infty}{\to} \int_\Lambda H(\lambda; x) \, d\mu(\lambda);$$

note that the limit is independent of the choice P_α.

Similarly, by (8.39),

$$\sum_k b_{\alpha k} = \int_\Lambda H_\alpha^{(2)}(\lambda; \infty) \, d\mu(\lambda) \underset{\alpha\to\infty}{\to} \int_\Lambda H^{(2)}(\lambda; \infty) d\mu(\lambda),$$

where $H^{(2)} = \lim_{\alpha\to\infty} H_\alpha^{(2)}$, completing the proof of (8.35). ∎

Note that the proof above holds if $\gamma_\delta(\nu)$ is replaced by

$$\gamma_\delta^1(\nu) = \int_{-\infty}^{\infty} \chi_1(x) d\nu_\delta(x),$$

where $\chi_1(x) = \chi(x) + O(x^2)$ $(x \to 0)$ is bounded. Thus we have

COROLLARY 8.8.1: *The statement of the Theorem holds if in (ii) γ_δ is replaced by γ_δ^1.*

§8.3 The R^*-Integral

Let (Λ, σ) be a metric space with σ-finite Borel measure μ. Let ϕ: $\mathbf{R}^+ \to \mathbf{R}^+$, and ρ_ϕ be the corresponding metric on $\mathcal{D}$. Let X: $\Lambda \to \mathcal{D}$ be ρ_ϕ-measurable, with a range consisting of independent r.v.'s. We now consider the general version of §3.1, i.e., the result of summing "samples" of X at an asymptotically dense set of points in Λ. In Chapter 9, Λ will be the spectrum of a von Neumann algebra of physical observables.

For $\epsilon > 0$, let $\Lambda_\epsilon = \{\lambda_{\epsilon j}\}_{j \in J_\epsilon} \subset \Lambda$ be at most countable. Note the elements of Λ need not be distinct. For $G \subset \Lambda$, let $N_\epsilon(G) = |\Lambda_\epsilon \cap G|$, where $|\cdot|$ denotes cardinality. We assume that (i) for any open set $G \subset \Lambda$,

$$\epsilon N_\epsilon(G) \underset{\epsilon\to 0}{\to} \mu(G), \tag{8.42a}$$

and (ii) for some $C < \mu(\Lambda)$, if B is an open ball with $\mu(B) \geq C$,

$$\epsilon N_\epsilon(B) \leq k\mu(B) \tag{8.42b}$$

for some fixed $k \in \mathbf{R}$. These conditions should be compared with (3.1).

DEFINITION 8.9: The net $\{\Lambda_\epsilon\}_{\epsilon>0}$ is a *μ-net* of points. We define

$$R^* \int_\Lambda X(\lambda)\phi(d\mu) \equiv \lim_{\epsilon\to 0} \phi(\epsilon)\sum_j X(\lambda_{\epsilon j}), \tag{8.43}$$

if the right hand limit (in law) is independent of the choice of $\{\Lambda_\epsilon\}_{\epsilon>0}$.

We proceed to relate the R^* to the Riemann integral.

DEFINITION 8.10: Let (Λ, σ) be a metric space, and $E \subset \Lambda$. Let $l = \operatorname{diam} E$, and $s \geq 0$ be the supremum of the diameters of all balls $B \subset E$. The pair (s, l) are the *dimensions* of E.

LEMMA 8.12.1: *Given $\epsilon > 0$, there is a partition P_α of Λ such that each $P_{\alpha k} \in P_\alpha$ has dimensions (s, l) where $s \geq \epsilon$, $l \leq 5\epsilon$. Furthermore, P_α can be chosen so that $\mu(\partial P_{\alpha k}) = 0$ for all k.*

Proof: Let $\mathcal{B}$ be a maximal set of disjoint ϵ-balls in Λ, and C be the collection of centers of $B \in \mathcal{B}$. For $B_\epsilon(\lambda) \in \mathcal{B}$, let $\tilde{B}(\lambda) \in \Lambda$ be the set of points closer to λ than to any other point in C. Let $\tilde{\mathcal{B}} = \{\tilde{B}(\lambda) : \lambda \in C\}$. If $\tilde{B} \in \tilde{\mathcal{B}}$, then the dimensions (s, l) of $\tilde{B}$ satisfy $(s, l) \subset (2\epsilon, 4\epsilon)$, i.e. $s \geq 2\epsilon$, $l \leq 4\epsilon$. There are at most a countable number of elements of $\tilde{\mathcal{B}}$ with non-null boundaries. Let $\tilde{B}_1, \tilde{B}_2, \ldots$ be an enumeration of this collection. For $\eta > 0$, $G \subset \Lambda$, let $B_\eta(G) = \{\lambda \in \Lambda : \sigma(\lambda, G) < \eta\}$. There exists $\eta_1 \leq \frac{\epsilon}{4}$ such that $B_{\eta_1}(\tilde{B}_1) = \{\lambda : \sigma(\lambda, \lambda_1) < \eta_1 \text{for some} \lambda_1 \in \tilde{B}_1\}$ has null boundary, since otherwise μ would not be σ-finite. We replace $\tilde{B}_1$ by $B_\eta(\tilde{B}_1)$ and decrease the remaining sets in $\tilde{\mathcal{B}}$ accordingly. We continue in this manner by then replacing $\tilde{B}_2$ by $B_{\eta_2}(\tilde{B}_2)$, $\eta_2 \leq \frac{\epsilon}{8}$, and, in general, replacing $\tilde{B}_k$ by $B_{\eta_k}(\tilde{B}_k)$, $\eta_k \leq \frac{\epsilon}{2^{k+1}}$, at each stage adjusting $\tilde{\mathcal{B}}$ as well. At the end of this process, all sets $\tilde{B} \in \tilde{\mathcal{B}}$ have null boundary, and dimensions $(s, l) \subset (\epsilon, 5\epsilon)$. We let $P_\alpha = \tilde{\mathcal{B}}$.∎

DEFINITIONS 8.11: A Borel measure on a metric space is *uniform* if for $\epsilon > 0$, every ball of radius ϵ has measure bounded below by some constant $c_\epsilon > 0$. A sequence $\{X_i\}$ of r.v.'s is *bounded* if the corresponding d.f.'s F_i satisfy $F \leq F_i \leq G$, where F and G are d.f.'s. It is *unbounded* if it is not bounded. A collection of sets is a *partial partition* of a set Λ if it satisfies all the requirements of a partition, except possibly $\bigcup_k P_k = \Lambda$. In particular, elements of Λ may be apportioned to several partition elements, if non-zero measures are divided correspondingly.

Note that a sequence X_n of r.v.'s is unbounded if for all $M \geq 0$ there exists $\epsilon > 0$ such that $\sup_{n\geq 1} P(|X_n| > M) \geq \epsilon$.

THEOREM 8.12: *Let (Λ, σ) be a metric space with uniform σ-finite Borel measure μ. If*

$X:\Lambda\to\mathcal{D}$ is Riemann integrable, then it is R^*-integrable, and the two integrals coincide.

Proof: Let $\{P'_\alpha\}_\alpha$ be an infinitesimal sequence of partitions of Λ, such that (see Lemma 8.12.1) $P'_{\alpha k}$ has dimensions $\dim P'_{\alpha k}\subset(\frac{1}{\alpha},\frac{5}{\alpha})$, and $\mu(\partial P'_{\alpha k})=0$. Since μ is uniform, there is a function $h(\alpha)>0$ such that $\mu(P'_{\alpha k})>h(\alpha)$. There exists a subpartition P_α of P'_α, consisting of sets with null boundary, with

$$h(\alpha)\le\mu(P_{\alpha k})\le 2h(\alpha). \tag{8.44}$$

If Λ is non-atomic, P_α can be constructed as a standard subpartition of P_α. If $P'_{\alpha k}$ has atoms we can write, according to previous conventions, $P'_{\alpha k}$ as a union of n copies of $P'_{\alpha k}$, each with measure $\frac{1}{n}\mu(P'_{\alpha k})$, with each a distinct element of P_α (this is a quick way of eliminating the problem of atoms, although the end result could be accomplished by subdividing only atoms).

Consider a μ-net $\Lambda'_\epsilon=\{\lambda'_{\epsilon j}\}_j$ satisfying

$$N'_\epsilon(P_{\alpha k})=\left[\frac{\beta\mu_{\alpha k}}{\epsilon}\right], \tag{8.45}$$

where $\beta>0$, $N'_\epsilon(P_{\alpha k})=|\Lambda'_\epsilon\cap P_{\alpha k}|$, and $[\cdot]$ is the greatest integer function. Given α and ϵ, subdivide $P_{\alpha k}$ into the partition $\{P_{\alpha ki}\}_i$, where each element satisfies $\mu_{\alpha ki}=\frac{\mu_{\alpha k}}{N'_\epsilon(P_{\alpha k})}$, where $\mu_{\alpha ki}=\mu(P_{\alpha ki})$, and $P_{\alpha ki}$ contains exactly one element $\lambda_{\alpha ki}\in\Lambda'_\epsilon$. If $\beta=1$, $X_{\alpha ki}=X(\lambda_{\alpha ki})$, then

$$\begin{aligned}\sum_{k,i}X_{\alpha ki}\phi(\mu_{\alpha ki})&=\sum_{k,i}X_{\alpha ki}\phi\left(\left[\frac{\mu_{\alpha k}}{\epsilon}\right]^{-1}\mu_{\alpha k}\right)\\&=\sum_{k,j}X(\lambda_{\epsilon j})\phi\left(\left[\frac{\mu_{\alpha k}}{\epsilon}\right]^{-1}\mu_{\alpha k}\right)\end{aligned} \tag{8.46}$$

where in the last sum k is a function of j defined by $\lambda_{\epsilon j}\in P_{\alpha k}$. If $\alpha=\alpha(\epsilon)$ increases to infinity sufficiently slowly as $\epsilon\to 0$, then by (8.45) and the Riemann integrability of X, if $X_{\epsilon j}=X(\lambda_{\epsilon j})$,

$$\sum_{k,j}X_{\epsilon j}\phi\left(\left[\frac{\mu_{\alpha k}}{\epsilon}\right]^{-1}\mu_{\alpha k}\right)\underset{\epsilon\to 0}{\Rightarrow}{}_R\int_\Lambda X\phi(d\mu). \tag{8.47}$$

Let $[\cdot]$ denote the greatest integer, α be fixed, $\{X_{\epsilon ki}\}_i$ be an indexing $\{X(\lambda_{\epsilon j}):\lambda_{\epsilon j}\in P_{\alpha k}\}$, and

$$a_{\epsilon k}=\phi\left(\left[\frac{\mu_{\alpha k}}{\epsilon}\right]^{-1}\mu_{\alpha k}\right)-\phi(\epsilon).$$

To prove that

$$\sum_{k,i}a_{\epsilon k}X_{\epsilon ki}\underset{\epsilon\to 0}{\Rightarrow}0, \tag{8.48}$$

we assume that it is not true. Recall that a space of finite measures with bounded total measure is compact in the topology of weak convergence. Thus there exists a sequence $\epsilon_\ell \underset{\ell\to\infty}{\to} 0$ such that (replacing ϵ_ℓ by ℓ) (*i*)

$$\sum_{k,i} a_{\ell k} X_{\ell k i} \underset{\ell\to\infty}{\Rightarrow} \nu \tag{8.49a}$$

for some $\nu \neq \delta_o$, or (*ii*)

$$\sum_{k,i} a_{\ell k} X_{\ell k i} \quad \text{is unbounded as } \ell \to \infty. \tag{8.49b}$$

We assume (*i*), since the argument is similar otherwise. By Theorems 7.13, 8.7, and our hypotheses, ϕ is homogeneous of positive order, so that $\frac{\phi^{-1}(a_{\ell k})}{\epsilon_\ell}$ is well-defined, and vanishes uniformly as $\ell \to \infty$.

Therefore

$$\left[\frac{\mu_{\alpha k}}{\epsilon_\ell}\right] \phi^{-1}(a_{\ell k}) \underset{\ell\to\infty}{\to} 0,$$

uniformly in k. Thus we may assume (by taking an ℓ-subsequence if necessary) that

$$\sum_\ell \left[\frac{\mu_{\alpha \ell}}{\epsilon_\ell}\right] \phi^{-1}(a_{\ell k}) \leq h(\alpha), \tag{8.50}$$

for each k. For $\ell = 1, 2, \ldots$, let $\{P^\ell_{\alpha k i}\}_i \equiv \mathcal{P}^\ell_{\alpha k}$ be a partial partition of $P_{\alpha k}$, into $\left[\frac{\mu_{\alpha k}}{\epsilon_\ell}\right]$ subsets, each of measure $\phi^{-1}(a_{\ell k})$, with $X_{\ell k i} \in P^\ell_{\alpha k i}$. By (8.50), $\mathcal{P}^\ell_{\alpha k}$ can be chosen so that $\mathcal{P}_{\alpha k} = \bigcup_\ell \mathcal{P}^\ell_{\alpha k}$ is a partial partition of $P_{\alpha k}$. By choosing an ℓ-subsequence, we may assume by (*i*) that

$$\rho^L\left(\sum_{i,k} a_{\ell k} X_{\ell k i}, \nu\right) \leq \frac{1}{C^\ell},$$

where $C > 0$ may be arbitrarily large. Therefore, the partial Riemann sum $\sum_{i,k,\ell} a_{\ell k} X_{\ell k i}$ becomes unbounded, contradicting Riemann integrability of X; this proves (8.48). Thus by (8.47),

$$\phi(\epsilon) \sum_j X_{\epsilon j} \underset{\epsilon\to 0}{\Rightarrow} R \int_\Lambda X \phi(d\mu) \tag{8.51}$$

By an exactly parallel argument, if in (8.45) $\beta = \beta(\alpha) \underset{\alpha\to\infty}{\to} 0$, and $\alpha = \alpha(\epsilon) \underset{\epsilon\to 0}{\to} 0$, the latter sufficiently slowly, then

$$\phi(\epsilon) \sum_j X_{\epsilon j} \underset{\epsilon\to 0}{\Rightarrow} 0. \tag{8.52}$$

Combining (8.51) and (8.52), if $\beta(\alpha) \underset{\alpha\to\infty}{\to} 0$, and

$$\left[\frac{(1-\beta)\mu_{\alpha k}}{\epsilon}\right] \leq N'_\epsilon(P_{\alpha k}) \leq \left[\frac{(1+\beta)\mu_{\alpha k}}{\epsilon}\right], \tag{8.53}$$

then

$$\phi(\epsilon)\sum X_{\epsilon j} \underset{\epsilon\to 0}{\Rightarrow}{}_R \int_\Lambda X\phi(d\mu). \tag{8.54}$$

Now consider a general μ-net $\Lambda_\epsilon = \{\lambda_{\epsilon j}\}_j$, not necessarily satisfying (8.53). Fix a partition $P_\alpha = \{P_{\alpha k}\}_k$ with null boundaries, satisfying (8.44). If two partition elements overlap (e.g., over atoms), elements of Λ_ϵ are apportioned among them alternatingly. By (8.42),

$$N_\epsilon(P_{\alpha k}) \equiv |\Lambda_\epsilon \cap P_{\alpha k}| \leq \frac{C}{\epsilon}$$

for some $C > 0$ independent of k and ϵ, while $\epsilon N_\epsilon(P_{\alpha k}) \underset{\epsilon\to 0}{\to} \mu_{\alpha k}$. Let

$$A_{\epsilon\alpha} = \left\{k : |\epsilon N_\epsilon(P_{\alpha k}) - \mu_{\alpha k}| \leq \frac{1}{\alpha}\right\}$$

and

$$\Lambda^{\epsilon,\alpha} = \bigcup_{k\in A_{\epsilon\alpha}} P_{\alpha k}. \tag{8.55}$$

Note that $\Lambda^{\epsilon,\alpha} \uparrow \Lambda$ as $\epsilon \to 0$.

We claim that if $S_\epsilon \subset \{j : \lambda_{\epsilon j} \notin \Lambda^{\epsilon,\alpha}\}$, then

$$\phi(\epsilon)\sum_{S_\epsilon} X_{\epsilon j} \underset{\epsilon\to 0}{\Rightarrow} 0. \tag{8.56}$$

The proof of this claim, briefly, follows by assuming the negation and forming an alternative, as in (8.49). A contradiction then follows along similar lines.

Let $\alpha = \alpha(\epsilon)$, and

$$\Lambda_*^{\epsilon,\alpha} = \bigcap_{\epsilon'\leq\epsilon} \Lambda^{\epsilon',\alpha(\epsilon')}.$$

Let $\alpha(\epsilon) \underset{\epsilon\to 0}{\to} 0$ sufficiently slowly that $\Lambda_*^{\epsilon,\alpha} \uparrow \Lambda$ as $\epsilon \to 0$, and

$$\phi(\epsilon)\sum_{S'_\epsilon} X_{\epsilon j} \Rightarrow 0, \tag{8.57}$$

if $S'_\epsilon \subset \{j : \lambda_{\epsilon j} \notin \Lambda_*^{\epsilon j}\}$; this is possible by (8.56). For $P_{\alpha k} \subset \Lambda_*^{\epsilon,\alpha}$

$$|\epsilon N_\epsilon(P_{\alpha k}) - \mu_{\alpha k}| \leq \frac{1}{\alpha}.$$

Let $\Lambda_\epsilon^1 = \{\lambda_{\epsilon j}^1\}_j \subset \Lambda$ be constructed so that $\Lambda_\epsilon^1 \cap P_{\alpha k} = \Lambda_\epsilon \cap P_{\alpha k}$ if $P_{\alpha k} \subset \Lambda_*^{\epsilon\alpha}$, and $\Lambda_\epsilon^1 \cap P_{\alpha k}$ differs from $\Lambda_\epsilon \cap P_{\alpha k}$ by the minimal number of elements such that

$$\left|\epsilon N_\epsilon^1(P_{\alpha k}) - \mu_{\alpha k}\right| \leq \frac{1}{\alpha} \tag{8.58}$$

for all k. Thus by (8.54),

$$\phi(\epsilon) \sum_j X(\lambda_{\epsilon j}^1) \underset{\epsilon \to 0}{\Rightarrow}_R \int_\Lambda X \phi(d\mu). \tag{8.59}$$

However, Λ_ϵ^1 is a small perturbation of Λ_ϵ, in that (8.57) and (8.59) imply that (8.59) holds also if $\lambda_{\epsilon j}^1$ is replaced by $\lambda_{\epsilon j}$. ∎

CHAPTER 9

JOINT DISTRIBUTIONS AND APPLICATIONS

We now use the mathematical machinery developed in the last two chapters to consider joint distributions of integrals of r.v.'s and apply them to statistical mechanics. The work in deriving joint distributions of quantum observables will, with the above formalism, be minimal.

§9.1 Integrals of Jointly Distributed R.V.'s

As in Chapter 8, let (Λ, σ) be a metric space with a σ-finite Borel measure μ. Assume that random vector-valued functions $\mathbf{X}(\lambda) = (X_1(\lambda), \ldots, X_n(\lambda))$ have a joint distribution for fixed $\lambda \in \Lambda$ and are independent for different λ. Let $\phi : \mathbf{R}^+ \to \mathbf{R}^+$ be given. Define the direct product $\mathcal{D}^n = \times_{i=1}^n \mathcal{D}$ and for $\vec{\nu}_i = (\nu_i^1, \nu_i^2, \ldots, \nu_i^n)$, $i = 1, 2, \ldots$, let

$$\rho_\phi^n(\vec{\nu}_1, \vec{\nu}_2) = \sum_{j=1}^n \rho_\phi(\nu_1^j, \nu_2^j)$$

(see eq. (7.21)). Define

$$\mathcal{R}_{\mathrm{ess}}(\mathbf{X}) = \{\vec{\nu} \in R(\mathbf{X}) : \mu(\mathbf{X}^{-1}(B_\epsilon(\nu))) > 0 \ \forall \epsilon > 0\}.$$

Let $\Phi_{\mathbf{X}}(\mathbf{t})$ denote the joint ch.f. of $\mathbf{X}$.

THEOREM 9.1: *The random vector $\mathbf{Y} = \int_\Lambda \mathbf{X}(\lambda)\phi(d\mu(\lambda))$ exists if and only if*
(i) for $\nu \in \mathcal{R}_{\mathrm{ess}}(\mathbf{X})$,

$$\psi_{\mathbf{X}}(\mathbf{t}) \equiv \lim_{\delta \to 0} \frac{1}{\delta}(u_{\mathbf{X}}(\phi(\delta)\mathbf{t}) - 1) \tag{9.1}$$

exists, and
(ii)

$$\psi(\mathbf{t}) \equiv \int_\Lambda \psi_{\mathbf{X}(\lambda)}(\mathbf{t})d\mu(\lambda) \tag{9.2}$$

converges absolutely. In this case, the ch.f. of $\mathbf{Y}$ is $e^{\psi(\mathbf{t})}$.

Proof: Suppose $\mathbf{Y}$ exists. If $\mathbf{a}$ is an n-vector and $\psi(\mathbf{t})$ is the log ch.f. of $\mathbf{Y}$, then for $t \in \mathbf{R}$, $\psi(t\mathbf{a})$ is the log ch.f. of $\mathbf{a}\cdot\mathbf{Y}$. By Theorem 8.5 the log ch.f of $\mathbf{a}\cdot\mathbf{Y} = \int_\Lambda \mathbf{a}\cdot\mathbf{X}(\lambda)\phi(d\mu(\lambda))$ is also given by $\int_\Lambda \psi_{\mathbf{a}\cdot\mathbf{X}(\lambda)}(t)\, d\mu(\lambda)$, adopting the notation of (9.1) for scalar r.v.'s. For a given $\lambda \in \Lambda$, if $\mathbf{X} \in \mathcal{R}_{\text{ess}}(X)$, then $\mathbf{a} \cdot \mathbf{X}(\lambda) \in \mathcal{R}_{\text{ess}}(\mathbf{a} \cdot \mathbf{X})$. Thus, since $\Phi_{\mathbf{a} \cdot \mathbf{X}}(t) = \Phi_{\mathbf{X}}(t\mathbf{a})$,

$$\lim_{\delta \to 0} \frac{1}{\delta}(\Phi_{\mathbf{a} \cdot \mathbf{X}}(\phi(\delta)t) - 1) = \lim_{\delta \to 0} \frac{1}{\delta}(\Phi_{\mathbf{X}}(\phi(\delta)t\mathbf{a}) - 1)$$

exists; since $\mathbf{a}$ is arbitrary, (i) follows. In addition, by Theorem 8.5,

$$\psi(t\mathbf{a}) = \int_{\Lambda} \psi_{\mathbf{a} \cdot \mathbf{X}(\lambda)}(t)\, d\mu(\lambda),$$

and the latter converges absolutely. Using $\psi_{\mathbf{a} \cdot \mathbf{X}}(t) = \psi_{\mathbf{X}}(t\mathbf{a})$ we conclude (9.2) converges absolutely, and $\mathbf{Y}$ has log ch.f $\psi(\mathbf{t})$, proving (ii).

Conversely, assume (i) and (ii) hold. Then given $\mathbf{a}$ and $X' \in \mathcal{R}_{\text{ess}}(\mathbf{a} \cdot \mathbf{X})$, there exists $\mathbf{X}' = (X'_1, \ldots, X'_n)$ such that $\mathbf{X}' \in \mathcal{R}_{\text{ess}}(\mathbf{X})$ and $\mathbf{a} \cdot \mathbf{X}' = X'$. Thus by (9.1) and (9.2) for $\psi_{\mathbf{X}'}$ and then by Theorem 8.5, $\mathbf{a} \cdot \mathbf{Y}$ exists. Since $\mathbf{a}$ was arbitrary, the proof is complete. ∎

§9.2 Abelian W^*-algebras

We present here a capsule summary of the spectral theory of W^*-algebras to be used in the next section. The material may be omitted without loss of continuity by those familiar with it.

Let $\mathcal{H}$ be a separable complex Hilbert space and $\mathcal{L}(\mathcal{H})$ be the bounded linear operators on $\mathcal{H}$. A W^* algebra $\mathcal{A}$ on $\mathcal{H}$ is an algebra of bounded operators on $\mathcal{H}$ closed under adjunction ($A \to A^*$), and closed in the weak operator topology on $\mathcal{H}$. A W^*-algebra is naturally a normed linear space with norm inherited from $\mathcal{L}(\mathcal{H})$. Let $\mathcal{A}^*$ be the space of bounded linear functionals on $\mathcal{A}$ as a Banach space and let the spectrum S of $\mathcal{A}$ consist of those $\phi \in \mathcal{A}^*$ which are also multiplicative, i.e., $\phi(A_1 A_2) = \phi(A_1)\phi(A_2)$, in the weak-star topology inherited from $\mathcal{A}^*$. The Gel'fand representation gives a canonical isometric algebraic star-isomorphism of $\mathcal{A}$ with the algebra (under multiplication) formed by the bounded continuous complex-valued functions $C_B(S)$ on S. This isomorphism is, for $a \in \mathcal{A}$,

$$a \leftrightarrow f \in C_B(S), \qquad \text{where } f(\phi) = \phi(a).$$

For $f \in C_B(S)$ let $T_f \in \mathcal{A}$ be the corresponding operator. For $x, y \in \mathcal{H}$, the map $f \to (T_f x, y)$ is non-negative and is bounded on $C_B(S)$ in its uniform (sup-norm) topology; hence there exists a Borel measure $\mu_{x,y}$ on S such that for $f \in C_B(S)$

$$(T_f x, y) = \int_S f\, d\mu_{x,y}. \tag{9.3}$$

DEFINITIONS 9.2: The measure $\mu_{x,y}$ is a *spectral measure.* A measure μ on S is *basic* if for any subset of S to be locally μ-null, it is necessary and sufficient that it be locally $\mu_{x,x}$-null for every $x \in \mathcal{H}$.

Clearly any two basic measures are absolutely continuous with respect to each other. We have (see [D2])

PROPOSITION 9.3: *If $\mathcal{H}$ is separable, then S carries a σ-finite basic measure.*

If μ is basic, then $C_B(S) = L^\infty(S,\mu)$. If Λ is a measure space, then $L^\infty(\Lambda)$, as a W^*-algebra acting on $L^2(\Lambda)$, is the *multiplication algebra* of Λ. We then have (see [D2]):

PROPOSITION 9.4: *Let μ be basic on S. Then the Gelfand isomorphism is the unique isometric star-isomorphism of the multiplication algebra $L^\infty(S,\mu)$ onto $\mathcal{A}$.*

If $\mathcal{A}$ is a W^*-algebra on $\mathcal{H}$, a possibly unbounded closed operator A is *affiliated* with $\mathcal{A}$, or $A \,\eta\, \mathcal{A}$, if A commutes with every unitary operator in the commutant $\mathcal{A}'$ of $\mathcal{A}$. If A is normal, then $A \,\eta\, \mathcal{A}$ if and only if $A = f(A_1)$, where $A_1 \in \mathcal{A}$, and $f : \mathbf{C} \to \mathbf{C}$ is Borel measurable.

A W^*-algebra is maximal abelian self-adjoint (masa) if it is properly contained in no other abelian W^*-algebra. In physics, a maximal commuting set of observables (that is, its spectral projections) generates a masa algebra. We require:

THEOREM 9.5 [Se1]: *Two masa algebras are algebraically isomorphic if and only if they are unitarily equivalent.*

DEFINITION 9.6: A measure space is *localizable* if every measurable set is a least upper bound of sets with finite measure in the partial order of set inclusion. Two measure spaces have isomorphic measure rings if there exists an algebraic isomorphism between their (Boolean) rings of measurable sets (modulo null sets). They are *isomorphic* if the isomorphism preserves measure.

THEOREM 9.7 [Se7]: *Two localizable measure spaces have isomorphic measure rings if and only if their multiplication algebras are algebraically star-isomorphic.*

§9.3 Computations with Value Functions

Although joint distributions of non-commuting observables in quantum statistical mechanics are difficult to describe (see [Se6] for the groundwork of such analysis), this is not the case with commuting observables, which are amenable to a standard (commutative) probabilistic analysis.

Let $\mathcal{A}$ be an abelian W^*-algebra, and let $A_i \geq 0$ and $A_i \,\eta\, \mathcal{A}$ $(1 \leq i \leq \ell)$. We assume that

A_i represent globally conserved quantites in a canonical ensemble whose density operator is formally

$$\rho = \frac{e^{-\overline{\beta}\cdot d\Gamma(\mathbf{A})}}{\operatorname{tr} e^{-\overline{\beta}\cdot\Gamma(\mathbf{A})}} \tag{9.4}$$

where $\overline{\beta} = (\beta_1, \ldots, \beta_\ell)$, an ℓ-tuple of positive numbers, $\mathbf{A} = (A_1, \ldots, A_\ell)$, and $d\Gamma(\mathbf{A}) = (d\Gamma(A_1), \ldots, d\Gamma(A_\ell))$ (Γ may be either Γ_B or Γ_F; see §1.2). Let $B_1, \ldots, B_n$ be self-adjoint operators affiliated with $\mathcal{A}$, whose joint distribution in the canonical ensemble of (9.4) is to be determined.

As with the single operators, ρ must be interpreted as a limit of density operators with discrete spectra. Proceeding in an analogous manner, let Λ be the spectrum of $\mathcal{A}$. Under Bose statistics, for $\lambda \in \Lambda$, let $\mathcal{N}_\lambda$ be the probability space $(\mathbf{Z}^+, \mathcal{B}_{\mathbf{Z}^+}, \mathcal{P}_\lambda)$ where $\mathbf{Z}^+ = \{0, 1, 2, \ldots\}$, $\mathcal{B}_{\mathbf{Z}^+}$ is its power set and for $z \in \mathbf{Z}^+$, $\mathcal{P}_\lambda(z) = e^{-\mathbf{A}(\lambda)\cdot\overline{\beta}z}(1 - e^{-\mathbf{A}(\lambda)\cdot\overline{\beta}})$, where $\mathbf{A}(\lambda) = (\lambda(A_1), \ldots, \lambda(A_\ell))$. Under Fermi statistics,

$$\mathcal{N}_\lambda = (\{0,1\}, \mathcal{B}_{\{0,1\}}, \mathcal{P}_\lambda),$$

where

$$\mathcal{P}_\lambda(0) = \frac{1}{1 + e^{-\overline{\beta}\cdot\mathbf{A}(\lambda)}}; \qquad \mathcal{P}_\lambda(1) = 1 - \mathcal{P}_\lambda(0). \tag{9.5}$$

Thus $\mathcal{N}_\lambda$ represents possible particle numbers in state λ. Spectral multiplicities need not be indicated here by duplication of spectral values, since they will be subsumed in a spectral measure μ. Let $\mathcal{N} = \prod_\lambda \mathcal{N}_\lambda$ be the direct product space with product measure $\mathcal{P} = \prod_\lambda \mathcal{P}_\lambda$.

DEFINITIONS 9.8: The pair $(\mathcal{N}, \mathcal{P})$ is the *canonical ensemble* over $\mathbf{A}$ at generalized inverse temperature $\overline{\beta}$ corresponding to the (generally formal) operator ρ.

Let N_λ be an r.v. on $\mathcal{N}$ defined by $N_\lambda(\prod_{\lambda' \in \Lambda} z_{\lambda'}) = z_\lambda$; N_λ is formally the number of particles in state λ. For $1 \le i \le n$ let

$$X_i(\lambda) \equiv \lambda(B_i) N_\lambda$$

be the r.v. on $\mathcal{N}$ representing the "total amount" of observable $d\Gamma(B_i)$ in state λ.

Our goal is to ascertain the joint distribution of $(d\Gamma(B_1), \ldots, d\Gamma(B_n)) \equiv d\Gamma(\mathbf{B})$; this is the distribution of a formal sum of $\mathbf{X}(\lambda) = (X_1(\lambda), \ldots, X_n(\lambda))$ over $\lambda \in \Lambda$. In order to study the distribution asymptotically, we first center $\mathbf{X}(\lambda)$. Let

$$\overline{\mathbf{X}}(\lambda) \equiv \mathbf{X}(\lambda) - \mathcal{E}(\mathbf{X}(\lambda)) \equiv \mathbf{B}(\lambda)\overline{N}_\lambda, \tag{9.6}$$

where $\overline{N}_\lambda \equiv N_\lambda - \mathcal{E}(N_\lambda)$, $\mathbf{B} = (B_1, \ldots, B_n)$.

Let μ be a (σ-finite) basic measure (see §9.2) on Λ. We study the Lebesgue integral $\int_\Lambda \overline{\mathbf{X}}(\lambda)(d\mu(\lambda))^{\frac{1}{2}}$. In the Bose case the r.v. N_λ is geometric with parameter $e^{-\overline{\beta}\cdot \mathbf{A}(\lambda)}$, and in the Fermi case it is Bernoulli with parameter given by (9.5). Let $\Psi_\lambda(t)$ be the ch.f. of $\overline{N}_\lambda$. If $\Phi_\lambda(\mathbf{t})$ is the ch.f. of $\overline{\mathbf{X}}(\lambda)$, then

$$\Phi_\lambda(\mathbf{t}) = \Psi_\lambda(\mathbf{t}\cdot \mathbf{B}(\lambda)). \tag{9.7}$$

Therefore, for ϕ small

$$\Phi(\phi \mathbf{t}) = 1 - \frac{1}{2}\phi^2 \mathbf{t}\mathbf{M}\mathbf{t}' \quad + \quad O(\phi^3),$$

where $M_{ij}(\lambda) = -\frac{\partial^2 \mu_\lambda}{\partial_{t_i}\partial_{t_j}}(0)$ is the covariance matrix of $\mathbf{X}$,

$$M_{ij}(\lambda) = \mathcal{E}(B_i(\lambda)B_j(\lambda)\overline{N}_{\lambda^2}) = B_i(\lambda)B_j(\lambda)\frac{e^{\overline{\beta}\cdot \mathbf{A}(\lambda)}}{\left(e^{\overline{\beta}\cdot \mathbf{A}(\lambda)} \mp 1\right)^2} \tag{9.8}$$

(see eq. (2.10)). The $-$ (+) holds for Bose (Fermi) statistics. Thus if $\phi(\delta) = \delta^{\frac{1}{2}}$,

$$\psi_\lambda(\mathbf{t}) \equiv \lim_{\delta\to 0}\left(\Phi_\lambda(\delta^{\frac{1}{2}}\mathbf{t}) - 1\right) = -\frac{1}{2}\mathbf{t}\mathbf{M}(\lambda)\mathbf{t}'.$$

Together with Theorem 9.1 this proves the reverse direction of

THEOREM 9.9: *If $\overline{\mathbf{X}}(\lambda)$ are the centered random vectors above, then the Lebesgue integral*

$$\mathbf{Y} = {}_L\int_\Lambda \overline{\mathbf{X}}(\lambda)(d\mu(\lambda))^{\frac{1}{2}} \tag{9.9}$$

exists if and only if

$$\mathbf{M} \equiv \int_\Lambda \mathbf{M}(\lambda)d\mu(\lambda) \tag{9.10}$$

converges absolutely component-wise. In this case $\mathbf{Y}$ *is normal, with covariance matrix* $\mathbf{M}$.

Proof: Only the forward direction remains. If (9.9) exists, then by Theorem 9.1,

$$\int_\Lambda \lim_{\delta\to 0}\frac{1}{\delta}(\Phi_\lambda(\delta^{\frac{1}{2}}\mathbf{t}) - 1)\ d\mu(\lambda)$$

converges absolutely for all $\mathbf{t}$. Thus, the same is true for (9.10). ∎

§9.4 Joint Asymptotic Distributions

By analogy with single operators, calculations like those above can be viewed as a direct treatment of an infinite volume limit. However, the algebra $\mathcal{A}$ and its spectrum Λ are actually limits of a net of discrete algebras $\mathcal{A}_\epsilon$ with spectra Λ_ϵ, and the joint distributions of $\mathbf{X}$ are asymptotic forms of those for discretized random vectors $\mathbf{X}_\epsilon$. In the discretized situation, all formal quantities, including the density operator, are well-defined.

Let μ be a σ-finite basic measure on Λ, and σ be a metric on Λ whose Borel sets $\mathcal{B}$ are just the μ-measurable sets (the existence of such metrics in applications will be shown explicitly; the integral, if it exists, is independent of the metric used). Let $\{\Lambda_\epsilon\}$ be a μ-net (Def. 8.9) of points in Λ, and for $A \in \mathcal{A}$, let $A_\epsilon = A\,|_{\Lambda_\epsilon}$.

DEFINITION 9.10: The algebra $\mathcal{A}_\epsilon = \{A_\epsilon : A \in \mathcal{A}\}$ is the *ϵ-discrete approximation* of $\mathcal{A}$ with respect to μ. Let $A_1, \ldots, A_\ell$ and $B_1, \ldots, B_n$ be as in §9.3.

In calculating R^*-integrals of r.v.'s $\mathbf{X}(\lambda) = \mathbf{B}(\lambda)N_\lambda$ we will be calculating $\epsilon \to 0$ limits of sums

$$\sum_{\lambda_{\epsilon i} \in \Lambda_\epsilon} \overline{\mathbf{X}}_\epsilon(\lambda_{\epsilon i}) \equiv \sum_{\lambda_{\epsilon i} \in \Lambda_\epsilon} \overline{\mathbf{X}}(\lambda_{\epsilon i}). \tag{9.11}$$

In order to use the machinery of §8.3, we need some assumptions about $\mathbf{A}(\lambda)$ and $\mathbf{B}(\lambda)$. Precisely, we require that $A_i(\lambda)$ be σ-continuous, and similarly for $B_i(\lambda)$ (that is, that there exist continuous representatives). We then have

THEOREM 9.11: *Let $\mathbf{A}(\lambda)$, $\mathbf{B}(\lambda)$, Λ, μ, and σ be as above, and μ be uniform (Def. 8.11) with respect to σ. If for some $\epsilon > 0$*

$$\int_\Lambda \sup_{\sigma(\lambda_1,\lambda) \leq \epsilon} |\mathbf{B}(\lambda)|^2 \frac{e^{\overline{\beta}\cdot\mathbf{A}(\lambda)}}{\left(e^{\overline{\beta}\cdot\mathbf{A}(\lambda)} \mp 1\right)^2} d\mu(\lambda) < \infty, \tag{9.12}$$

then the normalized distribution of $\mathbf{B}$ in the canonical ensemble over $\mathbf{A}$ at generalized inverse temperature $\overline{\beta}$ is jointly normal, with covariance

$$M_{ij} = \int_\Lambda B_i(\lambda)B_j(\lambda) \frac{e^{-\overline{\beta}\cdot\mathbf{A}(\lambda)}}{\left(e^{\overline{\beta}\cdot\mathbf{A}(\lambda)} \mp 1\right)^2} d\mu(\lambda), \tag{9.13}$$

where $-$ $(+)$ refers to Bose (Fermi) statistics. That is,

$${}_{R^*}\!\int_\Lambda \mathbf{B}(\lambda)\overline{N}_\lambda \,(d\mu(\lambda))^{1/2}$$

exists and is normal with covariance M_{ij}.

Proof: By Theorems 8.12, 8.7, and 9.9, we need only show that $\overline{\mathbf{X}}(\lambda) = \mathbf{B}(\lambda)\overline{N}_\lambda$ is Riemann integrable. By Corollary 8.8.1, it suffices to show

$$m^i_{\epsilon,\delta}(\lambda) \equiv \sup_{\substack{\delta_1 \leq \delta \\ \sigma(\lambda_1,\lambda)\leq\epsilon}} \frac{1}{\delta_1}\mathcal{E}\left(\theta(\sqrt{\delta_1}\,\overline{X}_i(\lambda)) + \chi_1(\sqrt{\delta_1}\,\overline{X}_i(\lambda))\right) \in L^1(\mu) \tag{9.14}$$

for some $\epsilon, \delta > 0$ and all $1 \leq i \leq n$, where

$$\chi_1(x) = \begin{cases} x; & x \leq 1 \\ 0; & x > 1. \end{cases}$$

Since $\mathcal{E}(\overline{X}_i(\lambda)) \equiv 0$, letting $I(x) = x$,

$$m^i_{\epsilon,\delta}(\lambda) \leq \sup_{\substack{\delta_1 \leq \delta \\ \sigma(\lambda_1,\lambda) \leq \epsilon}} \frac{1}{\delta_1}\mathcal{E}\Big(\delta_1 \overline{X}_i^2 + |(I - \chi_1)(\sqrt{\delta_1}\overline{X}_i)|\Big)$$

$$\leq c \sup_{\substack{\delta_1 \leq \delta \\ \sigma(\lambda_1 \leq \lambda) \leq \epsilon}} \frac{1}{\delta_1}\mathcal{E}(\delta_1 \overline{X}_i^2) = cB_i^2(\lambda)\mathcal{V}(\overline{N}_\lambda)$$

$$= \sup_{\sigma(\lambda_1,\lambda) \leq \epsilon} cB_i^2(\lambda)\frac{e^{\overline{\beta}\cdot\mathbf{A}(\lambda)}}{e^{\overline{\beta}\cdot\mathbf{A}(\lambda)} \mp 1}.$$

This shows that (9.14) is implied by (9.12), and completes the proof.∎

§9.5 Applications

An advantage of Einstein space over canonical spatially cut-off versions of Minkowski space is its possession of the full conformal group of summetries. In particular, the rotation group acts on Einstein space. Untractable (non trace-class) expressions involving generators of the conformal group in Minkowski space become tractable in Einstein space. For the purpose of evaluating joint distributions of observables Minkowski 4-space should be viewed as an infinite volume limit of Einstein space.

Theorem 9.11 allows evaluation of joint distributions in canonical ensembles (9.4), viewed as limits (according to a spectral measure μ) of systems with discrete spectrum. This section provides two explicit calculations.

(i) Non-vanishing chemical potential

Let S be a system with chemical potential $\mu > 0$, single particle Hamiltonian A, and formal density operator

$$\rho = \frac{e^{-\overline{\beta}\cdot d\Gamma(\mathbf{A})}}{\operatorname{tr} e^{-\overline{\beta}\cdot d\Gamma(\mathbf{A})}} \tag{9.15}$$

where $\overline{\beta} = (\beta, \mu)$, and $\mathbf{A} = (A, I)$. Note that $N = d\Gamma(I)$ is the particle number operator. This models an ensemble in which creation of particles requires energy μ. In this case the W^* algebra $\mathcal{A}$ generated by spectral projections of A and I is the bounded Borel functions of A.

If S consists of non-relativistic particles in Minkowski $n + 1$–space, the spectrum of $\mathcal{A}$ is measure theoretically equivalent to $\mathbf{R}^+$. The appropriate spectral measure is $dm =$

$\frac{n\pi^{n/2}}{\Gamma(\frac{n+2}{2})}E^{n-1}dE$ (see (6.12)). The joint distribution of $H = d\Gamma(A)$ and N is obtained by integrating $\mathbf{X} = \mathbf{B}(E)\overline{N}_E$, where $\mathbf{B}(E) = (E, 1)$, and $\overline{N}_E$ is a centered geometric r.v. with parameter $e^{-\beta E-\mu}$ (under Bose statistics). According to Theorem 9.11, the normalized joint asymptotic distribution of H and N is normal with covariance

$$\mathbf{M} = {}_{R^*}\int_\Lambda \begin{pmatrix} E^2 & E \\ E & 1 \end{pmatrix} \frac{e^{\beta E+\mu}}{(e^{\beta E+\mu} \mp 1)^2} dm.$$

Defining the generalized zeta function

$$\varsigma(n, x) \equiv \sum_{k=1}^{\infty} k^{-n} x^k \tag{9.16}$$

and using

$$\int_0^\infty \frac{x^n e^{x+\mu}}{(e^{x+\mu} \mp 1)^2} dx = \pm n!\, \varsigma(n, \pm e^{-\mu}) \qquad (n \geq 2), \tag{9.17}$$

we get

$$\mathbf{M} = \frac{\pm n! \pi^{\frac{n}{2}}}{\beta^n \Gamma(\frac{n+2}{2})} \begin{pmatrix} \frac{n(n+1)}{\beta^2}\varsigma(n+1, \pm e^{-\mu}) & \frac{n}{\beta}\varsigma(n, \pm e^{-\mu}) \\ \frac{n}{\beta}\varsigma(n, \pm e^{-\mu}) & \varsigma(n-1, \pm e^{-\mu}) \end{pmatrix} \qquad (n \geq 3)$$

Note that M_{11} and M_{22} coincides with (6.25) and the right side of (6.21), respectively, in the Bose case if $\mu = 0$. The calculation for $n < 3$ is similar, and thus omitted.

(*ii*) Density operator involving angular momentum (see [JKS])

In this model, the formal density operator in a system S is given by

$$\rho = \frac{e^{-\beta H - \gamma L^2}}{\operatorname{tr} e^{-\beta H - \gamma L^2}}; \tag{9.18}$$

where H and L^2 are energy and angular momentum [JKS, Se8]; we ignore chemical potential for simplicity. We will study S in Minkowski four-space M^4, as an infinite volume limit of Einstein space.

We remark first on the appropriate measure space of Theorem 9.11. Let $\mathcal{A}^*$ be the W^*-algebra generated by A and m^2, the single particle energy and angular momentum operators. Let Λ^* be the spectrum of $\mathcal{A}^*$, and μ^* be a σ-finite basic measure on Λ^*. By Proposition 9.4, $L^\infty(\mu^*)$ is star-isomorphic to $\mathcal{A}^*$.

Let $\Lambda = \mathbf{R}^+ \times \mathbf{Z}^+$, and $\mu = m \times c$, with m Lebesgue measure, and $c\{l\} = 2l + 1$, for $l \in \mathbf{Z}^+$. Then $\mathcal{A} = L^\infty(\mu)$ is star-isomorphic to $\mathcal{A}^*$, specifically through the correspondence

$$A \leftrightarrow M_E; \qquad m^2 \leftrightarrow M_{l(l+1)} \tag{9.19}$$

where E and l denote independent variables on $\mathbf{R}^+$ and $\mathbf{Z}^+$, respectively, and M denotes multiplication. Thus $L^\infty(\mu^*)$ and $L^\infty(\mu)$ are star-isomorphic. By Theorem 9.7, therefore, (Λ^*, μ^*) and (Λ, μ) have isomorphic measure rings, so that the image (under the isomorphism) of μ on Λ^* is equivalent to μ^*, and thus itself a basic measure.

Thus, henceforth we may restrict attention to (Λ, μ); μ is a physically appropriate measure on Λ, since it incorporates the asymptotics of the joint spectrum of A_ϵ and m_ϵ^2 in Einstein space U^4 of radius R proportional to $\frac{1}{\epsilon}$, as R becomes infinite. Specifically (see §6.3), the joint spectrum of A_ϵ and m_ϵ^2 in U^4 is

$$\{(\epsilon n, l(l+1)) : n, l \in \mathbf{Z}^+\},$$

this being the joint range of the corresponding functions on Λ.

The object of interest in studying the $\epsilon \to 0$ asymptotics of the joint distribution (covariance) of H and L^2 is, according to Theorem 9.11, the local covariance of $E\overline{N}_\lambda$ and $l(l+1)\overline{N}_\lambda$, where $\lambda = (E, l) \in \Lambda$, and $\overline{N}_\lambda$ is a centered geometric r.v. with parameter $e^{-\overline{\beta}\cdot\mathbf{A}(\lambda)}$, with $\overline{\beta} = (\beta, \gamma)$ and $\mathbf{A}(\lambda) = (E, l(l+1))$. This is the dyadic matrix

$$\mathbf{M}(\lambda) = \begin{pmatrix} E \\ l(l+1) \end{pmatrix} (E \quad l(l+1)) \mathcal{V}(N_\lambda) = \begin{pmatrix} E^2 & El(l+1) \\ El(l+1) & l^2(l+1)^2 \end{pmatrix} \frac{e^{\beta\cdot\mathbf{A}}}{(e^{\beta\cdot\mathbf{A}} \mp 1)^2}.$$

Integrating over Λ, we obtain as the covariance of H and L^2 in Minkowski space:

$$\mathbf{M} = \frac{1}{\beta} \begin{pmatrix} \frac{2}{\beta^2} & -\frac{1}{\beta}\frac{\partial}{\partial\gamma} \\ -\frac{1}{\beta}\frac{\partial}{\partial\gamma} & \frac{\partial^2}{\partial\gamma^2} \end{pmatrix} \xi_\mp(2, \gamma), \tag{9.20}$$

where

$$\xi_\mp(n, \gamma) = \pm \sum_{l=0}^{\infty} (2l+1)\, \varsigma(n, \pm e^{-\gamma l(l+1)}) \tag{9.21}$$

with ς given by (9.16). Note that, as indicated by the $\gamma \to 0$ limit in (9.21), the joint distribution of energy and angular momentum when $\gamma = 0$ is singular, with the conditional expectation of angular momentum infinite for every energy value in Minkowski space. This is to be expected, since for a fixed energy value the range of the angular momentum becomes unbounded as $R \to \infty$. See [JKS, Se8]) for an application of (9.20) to a model for the influence of angular momentum on the cosmic background radiation.

EPILOGUE

I would like to leave the reader by briefly identifying two significant open questions which arise in the present context.

The first is, what asymptotic probability distributions arise in a system whose spectral measure $d\mu \sim x^\alpha dx$ $(\alpha > 0)$ fails to have three continuous derivatives near 0 (Def. 3.11)? The failure to answer this question in Chapters 4 and 5 seems technical.

Second, how can these results be extended to a non-commutative setting (i.e., one involving non-commuting observables)? It seems that gage spaces, the non-commutative analogs of probability spaces [Se6], are the appropriate framework. This question may have very significant mathematical ramifications.

References

[AS] Abramowitz, M. and I. Stegun, *Handbook of Mathematical Functions,* U.S. Printing Office, Washington, D.C. 1968.

[BD] Bjorken, J.D. and S.D. Drell, *Relativistic Quantum Fields,* McGraw-Hill, New York, 1965.

[BDK] Bretagnolle, J., D. Dacunha Castelle, and J. Krivine, Lois stables et espaces L^p, *Ann. Inst. H. Poincaré* 2 (1966), 231-259.

[Ch] Chung, K.L., *A Course in Probability Theory,* Academic Press, New York, 1974.

[Che] Chentsov, N.N., Lévy-type Brownian motion for several parameters and generalized white noise, *Theor. Probability Appl.*2 (1957).

[D1] Dixmier, J., *C^*-Algebras,* North Holland, Amsterdam, 1977.

[D2] Dixmier, J., *Von Neumann Algebras,* North Holland, Amsterdam, 1981.

[GS] Gel'fand, I.M. and I.S. Sargsjan, *Introduction to Spectral Theory,* A.M.S., Providence, 1975.

[GV] Gel'fand, I.M. and N.J. Vilenkin, *Generalized Functions,* vol. 4. *Some Applications of Harmonic Analysis,* Academic Press, New York, 1964.

[GK] Gnedenko, B.V., and A.N. Kolmogorov, *Limit Distributions for Sums of Independent Random Variables,* Addison-Wesley, Reading, 1968.

[GR] Gradshteyn, I.S. and M. Ryzhik, *Table of Integrals, Series, and Products,* Academic Press, New York, 1965.

[G] Gumbel, E.J., *Statistics of Extremes,* Columbia University Press, New York, 1958.

[H] Hörmander, L., The spectral function of an elliptic operator, *Acta Math.* 121 (1968), 193-218.

[KL] Khinchin, A.Y., and P. Lévy, Sur les lois stables, *C.R. Acad. Sci. Paris* 202 (1936), 374-376

[JKS] Jakobsen, H., M. Kon, and I.E. Segal, Angular momentum of the cosmic background radiation, *Phys. Rev. Letters* 42 (1979), 1788-1791.

[Kh] Khinchin, A.Y., *Mathematical Foundations of Quantum Statistics,* Graylock Press, Albany, N.Y., 1960.

[La] Lamperti, J., Semi-stable stochastic processes, *Trans. Am. Math. Soc.* 104 (1962), 62-78.

[LL] Landau, L.D. and E.M. Lifschitz, *Statistical Physics,* Addison-Wesley, Reading, Mass.,

1958.

[Lé] Lévy, P., A special problem of Brownian motion and a general theory of Gaussian random functions, in *Proceedings of the Third Berkeley Symposium on Mathematical and Statistical Probability*, 1956.

[Lo] Loève, M., *Probability Theory, I,* Springer-Verlag, New York, 1977.

[Mc] McKean Jr., H.P., Brownian motion with several dimensional time, *Teor. Verojatnost. i Primenen* 8 (1963), 357-378.

[M] Malchan, G.M., Characterization of Gaussian fields with the Markov property, *Soviet Math. Dokl.* 12 (1971).

[R1] Rozanov, Ju. A., *Markov Random Fields*, Springer-Verlag, New York, 1982.

[R2] Rozanov, Ju. A., On Markovian fields and stochastic equations, *Mat. Sb.*106(148) (1978), 106-116.

[See] Seeley, R., An estimate near the boundary for the spectral function of the Laplace operator, *Am. J. Math.* 102 (1980), 869-902.

[Se1] Segal, I.E., *Decompositions of Operator Algebras. II: Multiplicity Theory,* American Mathematical Society Memoirs 9 (1951), 1-66.

[Se2] Segal, I.E., *Mathematical Cosmology and Extragalactic Astronomy,* Academic Press, New York, 1976.

[Se3] Segal, I.E., Tensor algebras over Hilbert spaces. I., *Trans. Am. Math. Soc.* 81 (1956), 106-134.

[Se4] Segal, I.E., Tensor algebras over Hilbert spaces. II., *Ann. Math.* 2 (1956), 60-175.

[Se5] Segal, I.E. and R.A. Kunze, *Integrals and Operators,* McGraw-Hill, New York, 1968.

[Se6] Segal, I.E., A noncommutative extension of abstract integration, *Ann. Math.* 57 (1953), 401-457.

[Se7] Segal, I.E., Equivalences of measure spaces, *Amer. J. Math.* 73 (1951), 275-313.

[Se8] Segal, I.E., Radiation in the Einstein universe and the cosmic background, *Phys. Rev. D* 28 (1983), 2393-2401.

[Si] Simon, B., Schrödinger semigroups, Bull. Am. Math Soc. 7 (1982), 445-526.

[V] Vanmarcke, E., *Random Fields: Analysis and Synthesis*, M.I.T. Press, Cambridge, U.S., 1983.

[WW] Whittaker, E.T. and G.N. Watson, *A Course of Modern Analysis,* Cambridge University Press, London, 1973.

INDEX

LIST OF SYMBOLS

Vol. 1062: J. Jost, Harmonic Maps Between Surfaces. X, 133 pages. 1984.

Vol. 1063: Orienting Polymers. Proceedings, 1983. Edited by J.L. Ericksen. VII, 166 pages. 1984.

Vol. 1064: Probability Measures on Groups VII. Proceedings, 1983. Edited by H. Heyer. X, 588 pages. 1984.

Vol. 1065: A. Cuyt, Padé Approximants for Operators: Theory and Applications. IX, 138 pages. 1984.

Vol. 1066: Numerical Analysis. Proceedings, 1983. Edited by D.F. Griffiths. XI, 275 pages. 1984.

Vol. 1067: Yasuo Okuyama, Absolute Summability of Fourier Series and Orthogonal Series. VI, 118 pages. 1984.

Vol. 1068: Number Theory, Noordwijkerhout 1983. Proceedings. Edited by H. Jager. V, 296 pages. 1984.

Vol. 1069: M. Kreck, Bordism of Diffeomorphisms and Related Topics. III, 144 pages. 1984.

Vol. 1070: Interpolation Spaces and Allied Topics in Analysis. Proceedings, 1983. Edited by M. Cwikel and J. Peetre. III, 239 pages. 1984.

Vol. 1071: Padé Approximation and its Applications, Bad Honnef 1983. Prodeedings. Edited by H. Werner and H.J. Bünger. VI, 264 pages. 1984.

Vol. 1072: F. Rothe, Global Solutions of Reaction-Diffusion Systems. V, 216 pages. 1984.

Vol. 1073: Graph Theory, Singapore 1983. Proceedings. Edited by K.M. Koh and H.P. Yap. XIII, 335 pages. 1984.

Vol. 1074: E.W. Stredulinsky, Weighted Inequalities and Degenerate Elliptic Partial Differential Equations. III, 143 pages. 1984.

Vol. 1075: H. Majima, Asymptotic Analysis for Integrable Connections with Irregular Singular Points. IX, 159 pages. 1984.

Vol. 1076: Infinite-Dimensional Systems. Proceedings, 1983. Edited by F. Kappel and W. Schappacher. VII, 278 pages. 1984.

Vol. 1077: Lie Group Representations III. Proceedings, 1982–1983. Edited by R. Herb, R. Johnson, R. Lipsman, J. Rosenberg. XI, 454 pages. 1984.

Vol. 1078: A.J.E.M. Janssen, P. van der Steen, Integration Theory. V, 224 pages. 1984.

Vol. 1079: W. Ruppert. Compact Semitopological Semigroups: An Intrinsic Theory. V, 260 pages. 1984

Vol. 1080: Probability Theory on Vector Spaces III. Proceedings, 1983. Edited by D. Szynal and A. Weron. V, 373 pages. 1984.

Vol. 1081: D. Benson, Modular Representation Theory: New Trends and Methods. XI, 231 pages. 1984.

Vol. 1082: C.-G. Schmidt, Arithmetik Abelscher Varietäten mit komplexer Multiplikation. X, 96 Seiten. 1984.

Vol. 1083: D. Bump, Automorphic Forms on GL (3,IR). XI, 184 pages. 1984.

Vol. 1084: D. Kletzing, Structure and Representations of Q-Groups. VI, 290 pages. 1984.

Vol. 1085: G.K. Immink, Asymptotics of Analytic Difference Equations. V, 134 pages. 1984.

Vol. 1086: Sensitivity of Functionals with Applications to Engineering Sciences. Proceedings, 1983. Edited by V. Komkov. V, 130 pages. 1984

Vol. 1087: W. Narkiewicz, Uniform Distribution of Sequences of Integers in Residue Classes. VIII, 125 pages. 1984.

Vol. 1088: A.V. Kakosyan, L.B. Klebanov, J.A. Melamed, Characterization of Distributions by the Method of Intensively Monotone Operators. X, 175 pages. 1984.

Vol. 1089: Measure Theory, Oberwolfach 1983. Proceedings. Edited by D. Kölzow and D. Maharam-Stone. XIII, 327 pages. 1984.

Vol. 1090: Differential Geometry of Submanifolds. Proceedings, 1984. Edited by K. Kenmotsu. VI, 132 pages. 1984.

Vol. 1091: Multifunctions and Integrands. Proceedings, 1983. Edited by G. Salinetti. V, 234 pages. 1984.

Vol. 1092: Complete Intersections. Seminar, 1983. Edited by S. Greco and R. Strano. VII, 299 pages. 1984.

Vol. 1093: A. Prestel, Lectures on Formally Real Fields. XI, 125 pages. 1984.

Vol. 1094: Analyse Complexe. Proceedings, 1983. Edité par E. Amar, R. Gay et Nguyen Thanh Van. IX, 184 pages. 1984.

Vol. 1095: Stochastic Analysis and Applications. Proceedings, 1983. Edited by A. Truman and D. Williams. V, 199 pages. 1984.

Vol. 1096: Théorie du Potentiel. Proceedings, 1983. Edité par G. Mokobodzki et D. Pinchon. IX, 601 pages. 1984.

Vol. 1097: R.M. Dudley, H. Kunita, F. Ledrappier, École d'Éte de Probabilités de Saint-Flour XII – 1982. Edité par P.L. Hennequin. X, 396 pages. 1984.

Vol. 1098: Groups – Korea 1983. Proceedings. Edited by A.C. Kim and B.H. Neumann. VII, 183 pages. 1984.

Vol. 1099: C.M. Ringel, Tame Algebras and Integral Quadratic Forms. XIII, 376 pages. 1984.

Vol. 1100: V. Ivrii, Precise Spectral Asymptotics for Elliptic Operators Acting in Fiberings over Manifolds with Boundary. V, 237 pages. 1984.

Vol. 1101: V. Cossart, J. Giraud, U. Orbanz, Resolution of Surface Singularities. VII, 132 pages. 1984.

Vol. 1102: A. Verona, Stratified Mappings – Structure and Triangulability. IX, 160 pages. 1984.

Vol. 1103: Models and Sets. Proceedings, Logic Colloquium, 1983, Part I. Edited by G.H. Müller and M.M. Richter. VIII, 484 pages. 1984.

Vol. 1104: Computation and Proof Theory. Proceedings, Logic Colloquium, 1983, Part II. Edited by M.M. Richter, E. Börger, W. Oberschelp, B. Schinzel and W. Thomas. VIII, 475 pages. 1984.

Vol. 1105: Rational Approximation and Interpolation. Proceedings, 1983. Edited by P.R. Graves-Morris, E.B. Saff and R.S. Varga. XII, 528 pages. 1984.

Vol. 1106: C.T. Chong, Techniques of Admissible Recursion Theory. IX, 214 pages. 1984.

Vol. 1107: Nonlinear Analysis and Optimization. Proceedings, 1982. Edited by C. Vinti. V, 224 pages. 1984.

Vol. 1108: Global Analysis – Studies and Applications I. Edited by Yu.G. Borisovich and Yu.E. Gliklikh. V, 301 pages. 1984.

Vol. 1109: Stochastic Aspects of Classical and Quantum Systems. Proceedings, 1983. Edited by S. Albeverio, P. Combe and M. Sirugue-Collin. IX, 227 pages. 1985.

Vol. 1110: R. Jajte, Strong Limit Theorems in Non-Commutative Probability. VI, 152 pages. 1985.

Vol. 1111: Arbeitstagung Bonn 1984. Proceedings. Edited by F. Hirzebruch, J. Schwermer and S. Suter. V, 481 pages. 1985.

Vol. 1112: Products of Conjugacy Classes in Groups. Edited by Z. Arad and M. Herzog. V, 244 pages. 1985.

Vol. 1113: P. Antosik, C. Swartz, Matrix Methods in Analysis. IV, 114 pages. 1985.

Vol. 1114: Zahlentheoretische Analysis. Seminar. Herausgegeben von E. Hlawka. V, 157 Seiten. 1985.

Vol. 1115: J. Moulin Ollagnier, Ergodic Theory and Statistical Mechanics. VI, 147 pages. 1985.

Vol. 1116: S. Stolz, Hochzusammenhängende Mannigfaltigkeiten und ihre Ränder. XXIII, 134 Seiten. 1985.